Theory and Technology of Mine Water Disaster Prevention and Resource Utilization

Theory and Technology of Mine Water Disaster Prevention and Resource Utilization

Helong Gu
Xueyi Shang
Huatao Zhao

MDPI

Basel • Beijing • Wuhan • Barcelona • Belgrade • Novi Sad • Cluj • Manchester

Editors

Helong Gu
College of Energy Engineering
Xi'an University of Science and Technology
Xian
China

Xueyi Shang
School of Resources and Safety Engineering
Chongqing University
Chongqing
China

Huatao Zhao
School of Geoscience and Technology
Southwest Petroleum University
Chengdu
China

Editorial Office
MDPI AG
Grosspeteranlage 5
4052 Basel, Switzerland

This is a reprint of articles from the Special Issue published online in the open access journal *Water* (ISSN 2073-4441) (available at: www.mdpi.com/journal/water/special_issues/V8126GD9HC).

For citation purposes, cite each article independently as indicated on the article page online and as indicated below:

Lastname, A.A.; Lastname, B.B. Article Title. *Journal Name* **Year**, *Volume Number*, Page Range.

ISBN 978-3-7258-2490-8 (Hbk)
ISBN 978-3-7258-2489-2 (PDF)
doi.org/10.3390/books978-3-7258-2489-2

Contents

About the Editors

Helong Gu

Dr. Gu Helong is an Associate Professor and master's supervisor. His main research interests lie in teaching and conducting research on coal mine dynamic disaster prevention and control, such as rock burst, high mining pressure, and coal and gas outburst. He has carried out extensive research on the theory and application of multi-field coupled coal-rock dynamic fracture mechanics in the field of rock dynamics. He has presided over six major projects, including the National Natural Science Foundation of China Youth Project, the China Postdoctoral Science Foundation Project, and the Yulin City Industry-University-Research Joint Project. Additionally, he has participated in five projects, including the National Key Research and Development Program. He has completed more than 30 horizontal projects focused on safe and efficient coal mining. As of now, he has published over twenty academic papers, among which there are nine SCI papers in top industry journals where he is the first author. Moreover, he has published three SCI papers in the top international journal the *International Journal of Rock Mechanics and Mining Sciences*. He has been authorized with more than 30 national invention patents, including 17 where he is the main inventor (top 2 in the list of inventors). He has formulated one industry standard, co-authored one textbook, and won two first-place awards and one third-place award in provincial and ministerial science and technology. Additionally, he has received one special contribution award from the Asian Rock Mechanics Conference. He serves as the Editor-in-Chief of a Special Issue for the international SCI journal *Water* (JCR second zone) and is a member of the Editorial Board of four well-known domestic and international journals, including the *Journal of Safety* and *Sustainability*. He is also the honorary Vice President of Technology for Shaanxi Zhongtai Energy Investment Co., Ltd.

Xueyi Shang

Dr. Xueyi Shang is an Associate Professor of Mining Engineering at Chongqing University, China. He holds both a Bachelor's and Ph.D. in Mining Engineering from Central South University. From 2017 to 2019, he was supported by the CSC to study at the Research School of Earth Sciences, Australia National University, Canberra, Australia. He is an expert in the Shanxi Province Coal Thousand Talents Think Tank and serves as a Youth Editorial Board Member for esteemed journals such as the *International Journal of Mining Science and Technology* and *Coal Science and Technology*.

Dr. Shang's research centers on mining methods, safety monitoring, and early warning. He has led a project funded by the National Natural Science Foundation of China, along with five other provincial- and ministerial-level projects, including the Special Postdoctoral Fund of China, and three enterprise-funded projects. He has also served as a Principal Investigator on four major projects, including key initiatives funded by the National Natural Science Foundation and the Chongqing Natural Science Foundation. His work has significantly advanced the development of efficient mining methods for deep fractured ore bodies, roof safety monitoring techniques, and high-precision microseismic inversion technology for predicting dynamic mining disasters. His contributions have earned him 6 provincial- and ministerial-level awards, 15 authorized invention patents, and the publication of over 30 SCI-indexed papers. He has also co-authored two textbooks.

Huatao Zhao

Zhao Huatao, Associate Professor, is a Young Expert of Sichuan Province and a Visiting Researcher at Kyoto University, Japan. He has long been dedicated to researching theories and technologies related to geological engineering and geological resources, covering areas such as rock mechanics, geological engineering, rock physics, and interdisciplinary research integrating artificial intelligence with multiple fields. He is a member of the Society of Exploration Geophysicists (SEG) and the Chinese Society for Rock Mechanics and Engineering, a Young Editorial Board Member of the *Journal of Safety and Sustainability*, and a reviewer for various journals, including *Tunnelling and Underground Space Technology*, *Scientific Reports*, the *Arabian Journal of Geosciences*, etc. He has received one provincial/ministerial-level First Prize for Scientific and Technological Progress, published over 20 papers, and holds more than 10 authorized national invention patents. He has led five national-, provincial-, and university-level research projects and has been the Principal Researcher on four major and general National Natural Science Foundation projects and three corporate innovation projects.

water

Editorial

Theory and Technology for the Prevention of Mine Water Disasters

Helong Gu [1,*], Xueyi Shang [2] and Huatao Zhao [3]

1 College of Energy Engineering, Xi'an University of Science and Technology, Xi'an 710054, China
2 School of Resources and Safety Engineering, Chongqing University, Chongqing 400044, China
3 School of Geoscience and Technology, Southwest Petroleum University, Chengdu 610500, China
* Correspondence: helonggu@xust.edu.cn

1. Introduction

The complex hydrogeological conditions in deep mining areas, coupled with large-scale and high-intensity mining activities, have triggered a series of mine water disasters and significant resource waste issues [1,2]. Mine water disaster prevention is crucial for the safe operation of mines. Studying the occurrence mechanism of water disasters, advanced detection, and disaster prevention can provide a theoretical basis and effective solutions for mine water treatment [3,4]. By studying fundamental mechanical theories of water-bearing coal rocks, technologies for the resource utilization of mine water, the mechanisms of surrounding rock, and advanced technologies for early warning and prevention of mine water disasters, accurate prediction and rapid response to mine water hazards can be achieved. This can minimize the losses caused by mine water disasters [5]. To ensure mine safety production, modern technological approaches are playing an increasingly vital role in both mine safety production and water resource management [6,7].

To address these challenges, scholars have made breakthroughs in various fields by leveraging modern technological approaches [8]. For example, they have employed a combination of methods to detect concealed water-bearing fractures and conducted in-depth studies on the geochemical characteristics of groundwater and mine water, providing technical support for a more accurate understanding of mine hydrogeological conditions and preventing water inrush [9]. Simultaneously, the application of advanced monitoring technologies and machine learning has enabled the identification of drainage tunnel intrusions and the prediction of water-bearing fracture expansions under thick, hard-roof conditions, providing effective means for timely detection and early warning in regard to mine safety disasters [10]. Research into the matching relationship between explosives and rock has improved blasting efficiency, reduced safety risks, and enhanced mine resource extraction efficiency [11,12]. In the Loess Plateau region, scholars have studied monitoring and assessment methods for landslides and debris flows, providing scientific theoretical guidance for disaster prevention and mitigation in the area [13]. By establishing coal mine water disaster risk assessment methods and researching the deformation and destruction characteristics and acoustic properties of sandstone at different water immersion depths, scholars have provided theoretical foundations for more precise identification and evaluation of mine water hazard risks and predicting the stability of water-inundated roadways, laying the groundwork for the development of effective prevention and control measures [14].

This Special Issue research explores the application of modern technology to enhance mine safety and promote the utilization of mine water resources from multiple perspectives. It provides critical theoretical foundations and technical support for ensuring mine safety and promoting sustainable development in the mining industry. The achievements will not only provide important theoretical support and technical backing for ensuring safe mine production but also offer powerful technological strength to promote sustainable development in the mining industry and to aid in building green mines.

Citation: Gu, H.; Shang, X.; Zhao, H. Theory and Technology for the Prevention of Mine Water Disasters. *Water* **2024**, *16*, 2952. https://doi.org/10.3390/w16202952

Received: 15 October 2024
Accepted: 16 October 2024
Published: 17 October 2024

2. An Overview of Published Articles

The incubation process, scientific evaluation, and efficient prevention of mine water disasters are of great significance for the safe and efficient mining of Chinese coal mines. Implementing efficient water detection, revealing the evolution mechanism of water disasters, and accurately evaluating the risk of water disasters are the essential technologies in regard to efficient prevention of water disasters. Ten papers in this Special Issue cover topics relevant to this subject.

Lu et al. proposed a geophysical–drilling–hydrochemical coupled method for the accurate detection of concealed water-conducting faults in coal mines. This method combines integrated mining geophysical exploration, hydrogeological drilling, and hydrochemical exploration (GDH). It conducts a thorough analysis of the background geological data obtained through surface exploration, enabling the prediction of potentially concealed water-conducting structures. Then, the seismic reflection method (SRM) and mine transient electromagnetic method (MTEM) are combined to detect the location and water-bearing properties of the target structures. Subsequently, the target drilling areas are defined by the anomalies detected by the integrated mine geophysical technique. Considering the hydrogeological information of water-conducting structures obtained through drilling methods, the hydrogeological information of water-conducting structures can be directly acquired. Through hydrochemical analysis, inrush water sources and their runoff conditions can be identified, and the spatial relationship between the source aquifers and mining space can be determined. Thus, the properties, scale, and configuration of the water-conducting structures can finally be evaluated. This multiple-method solution can accurately detect concealed water-conducting structures and help prevent water inrush disasters.

Yuan et al. developed a monitoring system to simulate intrusions in drainage tunnels based on distributed acoustic sensing (DAS) and the obtained typical characteristics of events such as rockfall events and water release. On this basis, an identification method for drainage tunnel intrusion events using principal component analysis (PCA) and neural networks was proposed. PCA reveals that amplitude-related parameters—amplitude, mean amplitude, and energy—significantly contribute to the classification of DAS signals, reducing the dimension of feature parameters by 54.8%. The accuracy of intrusion event classification is improved with PCA-processed data compared to unprocessed data, with overall accuracy rates of 79.1% for rockfall events and 72.7% for water release events. Additionally, the artificial neural network model outperforms the Bayesian and logistic regression models, demonstrating that an ANN has advantages in handling complex models for intrusion event classification.

Bai et al. investigated the initiation mechanism of landslide and debris flow disasters and collected their spreading patterns. Historical satellite images in the Laolang gully were used to generate three-dimensional topographic and geomorphological maps. Typical landslides were selected for measuring landslide thickness using a standard penetrometer and high-density electrical method. Numerical models were established to simulate the occurrence and development of landslides under different working conditions and to evaluate the spreading range based on the propagation algorithm and friction law. The results provide scientific and theoretical guidance for the prevention and control of rainfall-induced landslide and debris flow disasters in the Loess Plateau area.

Wang et al. examined the hydrogeochemical characteristics and evolution of Ordovician groundwater in the Zhuozishan coal mine in the northwest region of China. A total of 34 groundwater samples were collected for hydrogeochemical analyses and investigation of their evolution processes with the help of a piper trilinear diagram, a Gibbs diagram, and an ion ratio diagram. The author analyzed the runoff changes by analyzing the variation in characteristic ion concentrations, including sodium (Na^+), potassium (K^+), bicarbonate (HCO_3^-), chloride (Cl^-), sulfate (SO_4^{2-}), calcium (Ca^{2+}), and magnesium (Mg^{2+}), from the supply area to the discharge area. It was indicated that the northern part of the coalfield is characterized by a geological structure that creates a retention area for groundwater, resulting in an unordered runoff process with a complex formation mechanism. The middle region is free from geological constraints that would alter the flow direction, thus simplifying the process of groundwater formation. In contrast, the southern area experiences an increase in strata depth and fault blocking, which creates a retention zone, thereby making the groundwater formation process more complex. This research contributes to the effective management of groundwater resources in this coalfield and other mining sites.

Wang et al. developed a prediction model of water-conduction fracture zones based on a back propagation (BP) neural network by integrating theoretical analysis, field measurements, and algorithmic advancements. Firstly, they determined the height index system of the water-conducting fracture zone through an overburden migration analysis and correlation tests. This system includes mining height, buried depth, dip angle, working face width, and overburden rock lithology, and it features five groups of characteristic parameters. Then, the 35 pairs of minefield-measured data were collected to establish the measured height data set of the water-conducting fracture zone. Secondly, a BP neural network prediction model and a traditional support vector regression (SVR) prediction model were constructed based on a Pytorch framework, and the models were trained and tested by selecting data sets. Thirdly, the optimal prediction model was determined by comparing the model with the empirical model and the multiple regression model of mining regulations for coal pillar maintenance and pressure in buildings, water bodies, railways, and main shafts. Finally, a typical mine was selected for application to verify the suitability of the optimal model. The research results provide technical guidance for similar fragile coal seams.

Yang et al. analyzed the hydrochemical characteristics of highly mineralized coal mine water in semi-arid regions in Northwest China and revealed the formation mechanism of the water. Their research collected a total of 38 groundwater and mine water samples to examine the hydrogeochemical characteristics of high-salinity mine water using Piper diagrams and Gibbs diagrams, as well as isotope analyses and ion ratio coefficients. Additionally, they put forward corresponding mine water treatment recommendations. This study concludes that the TDS content of groundwater increases with hydrographic depth. The average TDS concentration of Quaternary, Luohe, and Anding groundwater is 336.87, 308.67, and 556.29 mg/L, respectively. However, the TDS concentration of Zhi Luo groundwater and mine water is 2768.57 and 3826.40 mg/L, respectively, which matches with high-salinity water. The hydrochemical type of Quaternary, Luohe, and Anding groundwater is predominantly the HCO_3-Ca type, while the hydrochemical type of Zhiluo groundwater and mine water is predominantly the SO_4-Na type. Furthermore, there are minimal differences observed in regard to δD and $\delta^{18}O$ values among these waters. It can be inferred that the Zhiluo formation in groundwater serves as the primary source of mine water supply, primarily influenced by the concentration processes caused by evaporation. The high salinity of mine water is closely related to the high salinity of Zhiluo groundwater. The high salinity of groundwater has gradually evolved under the control of the concentration caused by evaporation and rock-weathering processes. The dissolution of salt rock, gypsum, and other minerals serves as the material basis for the formation of high-salinity groundwater. In addition, the evolution of major ions is also affected by cation exchange.

Wei et al. take the 2102 fully mechanized mining face in the Balasu coal mine as the research background. By combining numerical simulation with physical simulation, the fracture evolution and strata movement characteristics in thick and hard overlying strata are simulated and analyzed and the formation mechanism of a water-conducting fracture in the overlying strata is revealed. It is further verified by field measurements of the development height of "two zones". The results show that the number and development height of fractures undergo the change–growth process of "slow–rapid–uniform". The water-conducting cracks in overlying strata show a dynamic expansion process of "local micro-cracks–jumping cracks–through cracks–water-conducting cracks". The fracture between the caving zone and fracture zone presents obvious layered characteristics. The overall shape of the water-conducting fracture zone is "saddle-shaped", and the maximum development height lags behind the coal mining face by about 180 m. This research provides theoretical guidance and a scientific basis for coal mine water disaster prevention under similar geological conditions.

Liu et al. studied the characteristics of deformation and damage as well as acoustic properties of sandstone in circular tunnel morphology under different inundation depths. In this study, the TAW-2000 press and static strain system was employed to investigate the mechanical properties, crack evolution, and deformation field distribution of sandstone with round holes at different immersion depths. Additionally, the study analyzed the impact of immersion depth on the characteristic parameters of acoustic emission. The results show that immersion depth is negatively correlated with the compressive strength and the modulus of elasticity of sandstone. As immersion depth increases, the duration of the compression and yield phases of the rock samples also increases, while the duration of the elastic phase remains relatively unchanged. These research findings have significant reference value for addressing issues related to water immersion and the degree of water saturation in roadways within rock engineering.

Zhu et al. studied the matching of explosives and rocks based on the hydrophysical properties of rocks. Firstly, they derived the theoretical solutions for crushing-zone energy, fragmentation energy, and fragment-throwing energy. Subsequently, concrete blocks with four types of cement–sand ratios were prepared and four types of emulsion explosives were used to conduct single-hole blasting tests. In these tests, a high-speed camera was employed to capture the trajectory of the blasting fragments, which were later collected. Finally, the crushing energy, fracturing energy, and fragment-throwing energy were calculated according to the test results and the basic parameters of the used explosives and concrete models. The results indicate that the size and distribution pattern of blasting blocks are significantly influenced by the hydrophysical properties of concrete and explosive properties. The higher the energy consumption in the rupture zone, the smaller the size of the fragments and the more uniform the distribution.

Xu et al. evaluated coal mine water hazard risk based on the combination of extension theory, game theory, and the Dempster–Shafer evidence theory. Firstly, they established a hierarchical water hazard risk evaluation index system. Then, matter–element theory in extension theory was used to establish a matter–element model for coal mine water hazard risk. The membership relationship between various evaluation indexes and risk grades of coal mine water hazard risk was quantified using correlation functions of the extension set theory, and the quantitative results were normalized to obtain basic belief assignments (BBAs) of risk grades for each index. Subsequently, the subjective weights of evaluation indexes were calculated using the order relation analysis (G1) method and the objective weights of evaluation indexes were calculated using the entropy weight (EW) method. The improved combination weighting method of game theory (ICWMGT) was introduced to determine the combination weight of each evaluation index, which was used to correct the BBAs of risk grades for each index. Finally, a fusion of the DS evidence theory, based on a matrix analysis, was used to fuse the BBAs, and the rating with the highest belief in the fusion result was taken as the final evaluation result. The evaluation model was applied to a water hazard risk evaluation of the Sangbei coal mine. The evaluation result was the II grade water hazard risk, which was in line with the actual engineering situation. The evaluation result was compared with the evaluation results of three methods, namely the expert scoring method, the fuzzy comprehensive evaluation method, and the extension method. Through this comparison, the scientificity and reliability of the method adopted in this paper were verified.

3. Conclusions

The research on the domains of geology and mining encompasses multiple facets, including rock mechanics, hydrogeology, and mining disaster prevention. By analyzing and synthesizing diverse research outcomes, a more lucid understanding of the overall developmental trends and crucial information in this field can be attained, providing a robust reference for subsequent research and engineering practices. The following is a summary of the conclusions from ten relevant articles.

(1) Uniaxial compressive tests on rock samples with circular holes demonstrate that, as the immersion depth increases, the uniaxial compressive strength and elastic modulus of sandstone notably decline, while the peak strain gradually rises. With a greater immersion depth, both the rapid decrease in uniaxial compressive strength and the increase in peak strain are more pronounced. When pressure is applied to rock samples at different immersion depths, water seepage leads to deviations in strain monitoring and the crack propagation and tensile failure points vary with the immersion depth.
(2) A numerical model is established to simulate the occurrence, development, and propagation range of debris flows under different working conditions. An appropriate DEM data resolution can reasonably assess debris flow risk, and a lower critical slope threshold results in more loose material flow, leading to wider and further landslide dispersion, which can simulate landslide movement under various conditions.
(3) Functional expressions for energy in various parts are derived based on principles like mass conservation. Different types of explosives and the physical conditions of water lead to different fragment sizes and distributions after an explosion, which provides important references for research on the matching of explosives with rocks in hydraulic engineering.
(4) Layers within the mining impact area exhibit a directional deflection in the stress chain, forming strong chain arches. Tensile stress concentration regions initially develop tensile cracks, and water-conducting fractures form through a specific process, reaching an approximate height of 108 m. The predictive model for extremely thin coal seams effectively forecasts the height of water-conducting fracture zones, while methods like BP neural networks can enhance predictive accuracy.
(5) High-salinity mine water can be treated using reverse osmosis combined processing techniques. The hydrogeochemical characteristics of Ordovician limestone groundwater are closely related to the flow direction, with three distinct hydrogeochemical regions being formed based on different flow mechanisms. Moreover, research on groundwater in mining areas indicates that a geophysical–drilling–hydrochemical coupled approach can accurately detect concealed water-conducting structures and that combining multiple methods can overcome the limitations of individual approaches, thus enhancing detection methods.
(6) Establishing a monitoring system based on DAS can capture typical event characteristics. The principal component analysis indicates significant correlations among feature parameters, showing higher classification accuracy after PCA processing, while artificial neural network models are excellent in handling complex models for classifying intrusion events.
(7) The risk assessment method based on extension theory, game theory, and DS evidence theory is proposed. An evaluation index system and risk assessment levels are established, optimizing the weighting result combinations to determine risk control factors and key prevention indicators. The assessment results are in good agreement with actual conditions and are scientifically reasonable, making them valuable for promotion and application.

In conclusion, these ten articles conduct in-depth research in the fields of geology and mining, covering rock properties, geological disasters, hydrogeology, and risk assessment. They reveal the changing patterns of rock mechanical properties under different conditions and the impact of explosions on rocks. Research on debris flows and landslides provides a foundation for disaster prevention and modeling, while groundwater studies present valuable information for treating high-salinity mine water and analyzing groundwater chemistry. The study of concealed water-conducting structure detection methods aids in enhancing safety in coal mining, and the proposed risk assessment approaches provide a sound decision-making basis for managing water hazards. These research findings not only enrich our theoretical knowledge on geology and mining but also offer practical engineering applications and methodological guidance.

Acknowledgments: As Guest Editor of the Special Issue "Theory and Technology of Mine Water Disaster Prevention and Resource Utilization", I would like to express my deep appreciation to all the authors whose valuable work was published in this issue and who have thus contributed to the success of this edition.

Conflicts of Interest: The authors declare no conflicts of interest.

List of Contributions:

1. Lu, T.; Liu, H.; Jia, H.; Wang, B. A Geophysical-Drilling-Hydrochemical Coupled Method for Accurate Detection of Concealed Water-Conducting Faults in Coal Mines. *Water* **2024**, *16*, 2619. https://doi.org/10.3390/w16182619
2. Liu, G.; Wang, S.;Wang, D.; Yang, Z.; Zan, Y. Characteristics of deformation and damage and acoustic properties of sandstone in circular tunnel morphology under varying inundation depths. *Water* **2024**, *16*, 2938. https://doi.org/10.3390/w16202938
3. Xu, X.; Wang, X.; Sun, G. Coal-Mine Water-Hazard Risk Evaluation Based on the Combination of Extension Theory, Game Theory, and Dempster-Shafer Evidence Theory. *Acoustics* **2024**, *4*, 2881. https://doi.org/10.3390/w16202881.
4. Zhu, Z.; Zhou, Z. Experimental Study for the Matching of Explosives and Rocks Based on Rock Hydrophysical Properties. *Water* **2024**, *16*, 1807. https://doi.org/10.3390/w16131807
5. Wei, D.; Gu, H.; Wang, C.; Wang, H.; Zhu, H.; Guo, Y. Extension Mechanism of Water-Conducting Cracks in the Thick and Hard Overlying Strata of Coal Mining Face. *Water* **2024**, *16*, 1883. https://doi.org/10.3390/w16131883
6. Wang, S.; Wang, T.; Yang, Z.; Tang, H.; Lv, H.; Xu, F.; Zhu, K.; Liu, Z. Hydrogeochemical Characteristics and Formation Processes of Ordovician Limestone Groundwater in Zhuozishan Coalfield, Northwest China. *Water* **2024**, *16*, 2398. https://doi.org/10.3390/w16172398
7. Yuan, P.; Zhang, W.; Shang, X.; Pu, Y. Intrusion event classification of drainage tunnel based on principal component analysis and neural network. *Water* **2024**, *16*, 2409. https://doi.org/10.3390/w16172409
8. Wan, B.; Bai, G.; An, N. Monitoring and Evaluation of Debris Flow Disaster in the Loess Plateau area of China: A Case Study. *Water* **2024**, *16*, 2539. https://doi.org/10.3390/w16172539
9. Wang, H.; Tian, J.; Li, L.; Chen, D.; Yuan, Y.; Li, B. Research on the development height prediction model of water conduction fracture zone under the condition of extremely thin coal seam mining. *Water* **2024**, *16*, 2273. https://doi.org/10.3390/w16162273
10. Yang, J.; Zhao, W.; Liang, X.; Xu, F. The Hydrochemical Characteristics and Formation Mechanism of Highly Mineralized Coal Mine Water in Semi-Arid Regions in Northwest China. *Water* **2024**, *16*, 2244. https://doi.org/10.3390/w16162244

References

1. Yang, Y.; Guo, T.; Jiao, W. Destruction processes of mining on water environment in the mining area combining isotopic and hydrochemical tracer. *Environ. Pollut.* **2018**, *237*, 356–365. [CrossRef]
2. Yin, L.; Ma, K.; Chen, J.; Xue, Y.; Wang, Z.; Cui, B. Mechanical model on water inrush assessment related to deep mining above multiple aquifers. *Mine Water Environ.* **2019**, *38*, 827–836. [CrossRef]
3. Khan, A.; Gupta, S.; Gupta, S. Multi-hazard disaster studies: Monitoring, detection, recovery, and management, based on emerging technologies and optimal techniques. *Int. J. Disaster Risk Reduct.* **2020**, *47*, 101642. [CrossRef]
4. Su, B.; Liu, S.; Deng, L.; Gardoni, P.; Krolczyk, G.; Li, Z. Monitoring direct current resistivity during coal mining process for underground water detection: An experimental case study. *IEEE Trans. Geosci. Remote Sens.* **2022**, *60*, 1–8. [CrossRef]
5. Zheng, Q.; Wang, C.; Zhu, Z. Research on the prediction of mine water inrush disasters based on multi-factor spatial game reconstruction. *Geomech. Geophys. Geo-Energy Geo-Resour.* **2024**, *10*, 41. [CrossRef]
6. More, K.; Wolkersdorfer, C.; Kang, N.; Elmaghraby, A. Automated measurement systems in mine water management and mine workings–A review of potential methods. *Water Resour. Ind.* **2020**, *24*, 100136. [CrossRef]
7. Dong, S.; Xu, B.; Yin, S.; Han, Y.; Zhang, X.; Dai, Z. Water resources utilization and protection in the coal mining area of northern China. *Sci. Rep.* **2019**, *9*, 1214. [CrossRef]
8. Zhang, S.; Wang, H.; He, X.; Guo, S.; Xia, Y.; Zhou, Y.; Yang, S. Research progress, problems and prospects of mine water treatment technology and resource utilization in China. *Crit. Rev. Environ. Sci. Technol.* **2020**, *50*, 331–383. [CrossRef]
9. Zhang, S.; Liu, Y. A simple and efficient way to detect the mining induced water-conducting fractured zone in overlying strata. *Energy Procedia* **2012**, *16*, 70–75. [CrossRef]

10. Lian, H.; Xu, B.; Tian, Z.; Liu, D.; Yang, Y.; Pan, G.; Wang, R. Design and implementation of mine water hazard monitoring and early warning platform. *Coal Geol. Explor.* **2021**, *49*, 198–207.
11. Xia, C.; Song, Z.; Tian, L.; Liu, H.; Lu, W.; Wu, X. Numerical analysis of effect of water on explosive wave propagation in tunnels and surrounding rock. *J. China Univ. Min. Technol.* **2007**, *17*, 368–371. [CrossRef]
12. Zhu, Z.; Zhou, Z. Experimental Study for the Matching of Explosives and Rocks Based on Rock Hydrophysical Properties. *Water* **2024**, *16*, 1807. [CrossRef]
13. Crosta, G.; Frattini, P. Rainfall-induced landslides and debris flows. *Hydrol. Process. Int. J.* **2008**, *22*, 473–477. [CrossRef]
14. Zhang, Q.; Tong, Z.; Shen, D.; Luo, Z.; Ding, W.; Xu, H. Mechanism investigation on water and mud inrush disasters when Wangjiazhai tunnel passing through the Tertiary water-rich sandstone strata. *Environ. Earth Sci.* **2024**, *83*, 479. [CrossRef]

water

MDPI

Article

A Geophysical-Drilling-Hydrochemical Coupled Method for Accurate Detection of Concealed Water-Conducting Faults in Coal Mines

Tuo Lu [1,*], Haodong Liu [1], Hailiang Jia [2] and Bo Wang [3]

[1] College of Geology and Environment, Xi'an University of Science and Technology, Xi'an 710054, China; 19591565095@163.com
[2] College of Architecture and Civil Engineering, Xi'an University of Science and Technology, Xi'an 710054, China; hailiang.jia@xust.edu.cn
[3] State Key Laboratory of Deep Geotechnical Mechanics and Underground Engineering, China University of Mining and Technology, Xuzhou 710054, China; wbsyes@126.com
* Correspondence: lutuoxust@163.com

Abstract: The detection of concealed water-conducting structures is essential for preventing water inrush disasters. Aiming to mitigate the limitations inherent in using any single technique, a comprehensive approach that combines integrated mining geophysical exploration, hydrogeological drilling, and hydrochemical exploration (GDH) is proposed for the exploration of concealed water-conducting structures. By conducting a thorough analysis of the background geological data obtained through surface exploration, potentially concealed water-conducting structures can be predicted. Then, a combination of the seismic reflection method (SRM) and mine transient electromagnetic method (MTEM) can be used to detect the location and water-bearing properties of the target structures. Afterwards, the target drilling areas are defined by the anomalies detected by the integrated mine geophysical technique, and the drilling method can directly acquire the hydrogeological information of water-conducting structures and verify the results of the geophysical methods. By means of hydrochemical analysis, inrush water sources and their runoff conditions can be identified, and the spatial relationship betweenof the source aquifers and mining space can be determined; hence, the properties, scale, and configuration of the water-conducting structures can finally be evaluated. Employing a water-conducting fault in a mine as a case study, we verified that the integrated method overcomes the limitations and possible biases of each method, providing a multiple-method solution that can accurately detect concealed water-conducting structures to help prevent water inrush disasters.

Keywords: concealed water-conducting structures; combined method for accurate detection; integrated mining geophysical prospecting; hydrogeological drilling; hydrochemical analysis

Citation: Lu, T.; Liu, H.; Jia, H.; Wang, B. A Geophysical-Drilling-Hydrochemical Coupled Method for Accurate Detection of Concealed Water-Conducting Faults in Coal Mines. *Water* **2024**, *16*, 2619. https://doi.org/10.3390/w16182619

Academic Editor: Andrzej Witkowski

Received: 25 August 2024
Revised: 10 September 2024
Accepted: 13 September 2024
Published: 15 September 2024

1. Introduction

China is the largest coal-producing country in the world. However, due to its extremely complex geological structural conditions, it is also one of the countries suffering from the most serious coal mine water disasters. Statistics indicate that from 2000 to 2023, a total of 1220 water hazard accidents took place in coal mines across China, leading to 5097 fatalities. As shown in Figure 1, over the past decade, China's coal output has been on a gradual rise, and mine water accidents have been effectively curbed. Nevertheless, there are still dozens of such accidents occurring every year. Particularly in the past five years, the number of fatalities caused by mine water accidents shows an upward trend [1], highlighting the significance of mine water disaster prevention.

Coal mine water inrush accidents are typically induced by the excavation and disclosure of unexpected concealed water-conducting structures, such as faults, fissure zones, and karst. For instance, 80% of floor water hazard accidents are associated with hidden water-bearing faults [2]. Upon disclosure during excavation or connecting by drilling, these structures may channel high-speed water toward the tunnel from various sources, causing disastrous consequences [3,4]. Such structures undermine the integrity and stability of the rock mass and regulate the distribution and inrush of water; thus, determining the accurate location and water-bearing state of concealed water-conducting structures is an essential precondition for preventing water disasters and implementing control measures [5–7].

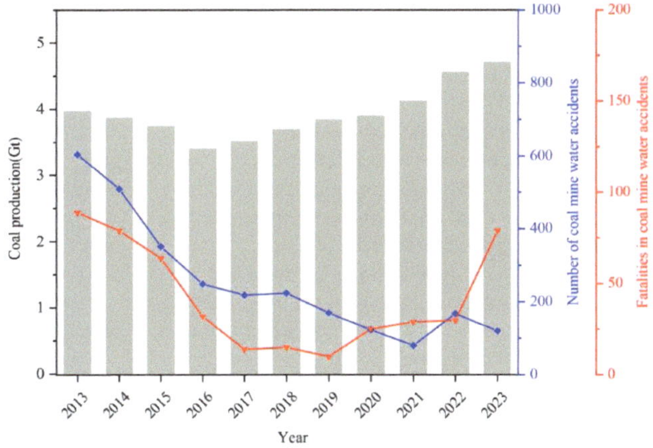

Figure 1. The coal production and the number of coal mine water accidents and fatalities from 2013 to 2023 in China. (The data are sourced from the National Bureau of Statistics and the National Mine Safety Supervision Administration of China).

At present, research on the detection of concealed geological structures mainly focuses on assessing and improving the effects of geophysical, geochemical, and drilling methods [8–10]. Considerable efforts have been made to detect concealed structures and estimate their water yield properties using the isotope tracing method, hydrochemical analysis [11,12], DC electrical method [5,13], mine transient electromagnetic method [14–16], seismic exploration [17–19], hydrogeological drilling, and borehole television. However, there are limitations and potential biases in interpreting water-conducting structures using a single method or technique. For instance, there is ambiguity in water source recognition, multiplicity in geological methods, and limited vision in drilling methods [20]. By taking advantage of each method, proposing a multimethod prospecting system becomes an inevitable choice. Unfortunately, such a multimethod prospecting system has not been well developed [21].

In the present study, through an in-depth analysis of the advantages and limitations of each method, a geophysical-drilling-geochemical coupled method for the accurate detection of concealed water-conducting faults in coal mines is first proposed. Then, the proposed method is applied to a real case of water-conducting fault detection in a mine in Xuzhou, China. At last, the limitations of the proposed method is are discussed.

2. Proposal of a Geophysical-Drilling-Geochemical Coupled Method

Exploration conducted on the surface has successfully identified the major geological structures within the mining area. However, smaller concealed structures often pose a greater threat to coal mine safety. These structures, which lack obvious surface exposure or distinguishing features, are typically small-scale formations within the rock and are characterized by high concealment and significant disaster-causing potential. Additionally, the detection of these concealed structures usually occurs concurrently with mining activities, necessitating strict accuracy and timeliness in exploration. The tasks involved in exploring these hidden structures are comprehensive, covering aspects such as the presence of water, water-conducting channels, and sources of water.

Geophysical exploration, drilling, and geochemical analysis are commonly used for advanced detection, with geophysical exploration and drilling being the most prominent. While these methods can partially address geological challenges, they each have significant limitations when used independently. Geophysical exploration relies on interpreting detected geophysical field characteristics; however, as an indirect method, it faces the issue of multiple solutions—different geological conditions can produce similar geophysical characteristics. Additionally, each geophysical technique has its own application constraints. For example, seismic methods offer high positioning accuracy but are not sensitive to water, whereas electrical methods are effective for water detection but have limited positioning accuracy due to the volume effect. Drilling, on the other hand, has a limited control range and blind spots in exploration. If a borehole does not reach a water-bearing body, it only confirms the absence of water at that specific location, necessitating multiple boreholes to accurately detect the water distribution across the entire mining face. This approach requires a long construction period and substantial investment. Geochemical exploration, particularly hydrochemical analysis, is often used to infer geological environments, water sources, and potential anomalies. However, it is a reactive method, often requiring contact with groundwater, which means that water has already surfaced or samples must be obtained through drilling. As a result, in coal mine water detection, geochemical exploration typically serves as a supplement to geophysical exploration and drilling. Therefore, no single method is sufficient for the refined detection of concealed water-conducting structures.

Geological structures have specific physical and chemical characteristics that distinguish them from the surrounding rocks. By employing a multimethod approach to study these properties, we can gather substantial information. Thus, we propose a combined method of geophysical exploration, drilling, and geochemical exploration (GDH). This method should follow a sequence of prediction, detection, verification, and identification steps to accurately detect concealed water-conducting structures. First, using basic geological information, we can predict the areas where these structures might develop. Then, using the integrated mining geophysical method, the location, boundary, and water yield properties of geological structures can be further detected. Directional drilling, guided by geophysical data, allows for the direct detection and verification of the structure's form and water characteristics. Finally, by utilizing strata information and identifying the source aquifer through geochemical exploration, we can estimate the properties and scale of the structures (Figure 2).

(1) Prediction of potential concealed water-conducting structures through data analysis aided by surface exploration

By thoroughly analyzing geological data, including mine strata, structures, and hydrology, we can examine the relationship between the main coal seam and the sedimentary structure of aquifers, as well as identify water sources and conduction channels. Evaluation methods are employed to assess the characteristics of water hazard threat zoning and predict suspicious areas surrounding the production area, especially in the vicinity of large geological structures interpreted by surface exploration and areas with low exploration accuracy. These zones may contain undetected or associated geological structures. When mining activities approach these areas, the detection of concealed geological structures is essential.

(2) Detection of the location and water abundance of water-conducing structures by integrated geophysical exploration (IGE)

As a deep-reaching and non-destructive technology, mine geophysical methods have been widely employed in the detection of geological structures and their water-bearing properties [22]. Geological structures often disrupt the continuity and integrity of the rock mass, leading to fractures and low cementation in tectonic belts. The lithology in these areas may be altered or water-filled, resulting in variations in resistivity and wave impedance (velocity and density). These anomalies in resistivity and wave impedance form the basis of electrical and seismic exploration in mining. Electrical methods (EM), including direct current methods and electromagnetic methods, are sensitive to the water-bearing properties of structures. However, some methods may magnify the volume of low-resistivity anomalies due to the volume effect [15]. However, the seismic method (SM), which employs the travel time characteristics of elastic waves, can detect the location and boundary of geological structures more accurately. Integrated geophysical exploration (IGE) involves adopting at least two underground geophysical exploration methods and taking advantage of the strengths of each method to complement and validate one another. As a result, IGE can provide reliable information on the formation of concealed geological structures and even enable the quantitative detection of water volume within these structures. This information converges in the target area and provides important data for planning hydrogeological drilling.

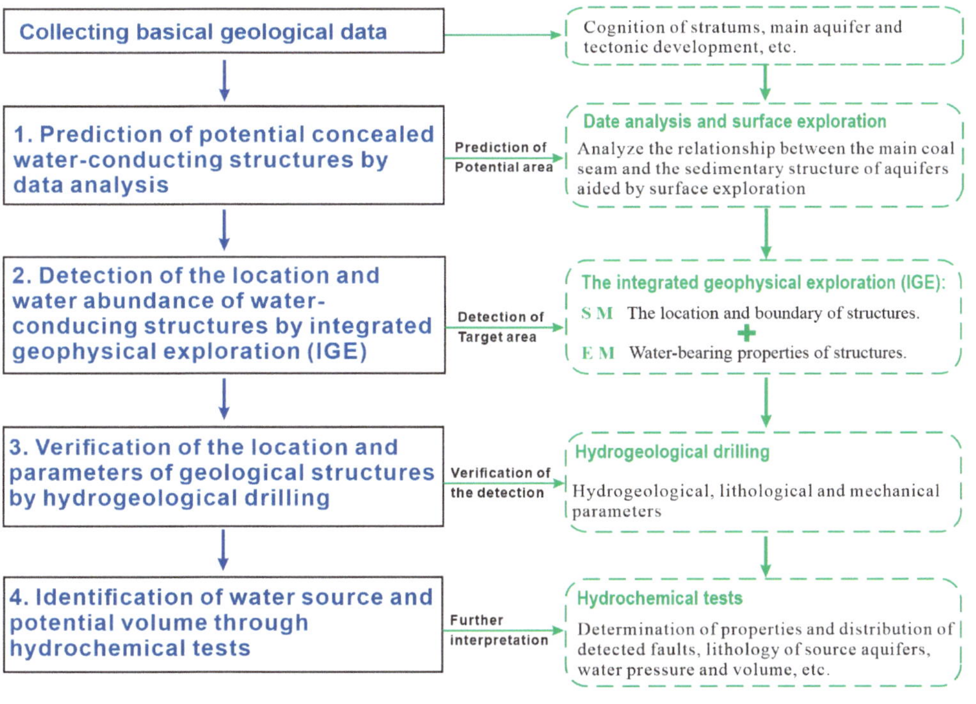

Figure 2. The Geophysical-drilling-geochemical (GDH) Coupled Method.

(3) Verification of the location and parameters of geological structures by hydrogeological drilling

Hydrogeological drilling is the most efficient method for water detection and drainage in mines; it is also the most direct way to obtain actual geological data and assess the mechanical characteristics of an unexplored area. By observing the water inflow in the drill hole, rock strength, core recovery, and changes in water yield, we can determine the lithology of the aquifer, distribution of geological structures, and water pressure, among other parameters. However, the information provided by drilling is limited to the area surrounding the drill hole, while geophysical exploration results can provide a targeted zone for drilling design and set a more relevant aim for drilling construction. Additionally, the analysis of drilling samples can serve to verify the results obtained from geophysical exploration.

(4) Identification of water source and potential volume through hydrochemical tests

The properties of a water source are crucial for detecting and managing concealed water-conducting structures. Underground water formation is influenced by various factors, including the sedimentary sequence, host lithology, geochemical environment, and geological structures[23]. The chemical components and hydrochemical characteristics of underground water vary in different aquifers. By analyzing hydrochemical data, we can identify the recharge source and runoff conditions of underground water samples, determine the spatial relationship between source aquifers and the mining area, and thereby deduce the properties, scale, and configuration of water-conducting structures. These results provide a foundation for subsequent water prevention and control measures.

3. Application of the GDH Method

3.1. Background Information

As shown in Figure 3, the study site is a coal mine located in the northern region of Xuzhou, China. The mining area lies within the coal-bearing strata of the Permian-Carboniferous system, with the primary coal seam being the No. 7 coal seam of the Shanxi Formation. The mine contains five main aquifers: a bottom aquifer of the Quaternary system, a sandstone aquifer on the roof of the No. 7 coal seam, a sandstone aquifer on the roof of the No. 8 coal seam, the No. 4 limestone aquifer of the Taiyuan Formation, and the Ordovician limestone aquifer. In Figure 3, only the sandstone aquifer in the roof of coal seam No. 7, the sandstone aquifer in the roof of coal seam No. 8, and limestone aquifer No. 4 are presented. Their rock names are displayed in blue. Among these, the No. 4 limestone aquifer of the Taiyuan Formation has the greatest impact on mining operations due to its thickness of up to 10 m and its proximity to the seam, which is approximately 66 m below the No. 7 coal seam. The Quaternary aquifer and the Ordovician limestone aquifer are farther from the mining area and generally have minimal impact on mining activities unless there is an extremely large fault.

Fractures, particularly tension normal faults, are the major geological structures in this mining area and are known for their significant water abundance and conductivity. Faults tend to reduce the thickness of the aquiclude and decrease the distance between the coal seam and aquifers. Additionally, tensile stress can enhance the development of fissures in limestone, which promotes water enrichment. According to the three-dimensional seismic data and actual exposure information from roadways, a normal fault with a displacement of approximately 50 m has been identified to the west of the shaft station (Figure 4). There may be unexpected associated faults ahead of the mining face. To ensure safe roadway tunneling, the proposed GDH method is applied to detect potential water-bearing faults.

3.2. Accurate Detection Using GDH Method

3.2.1. Integrated Geophysical Method

(1) Mine seismic exploration

To determine the geological conditions ahead of the roadway, the seismic reflection wave method (SRM) and the mine transient electromagnetic method (MTEM) are employed. The seismic survey line was placed on the right side of the roadway, and the measurement points are shown in Figure 5. A 12-pound copper hammer served as the source, and velocity geophones with a natural frequency of 100 Hz were used as the receivers. The intervals between the geophones and sources were both 2 m.

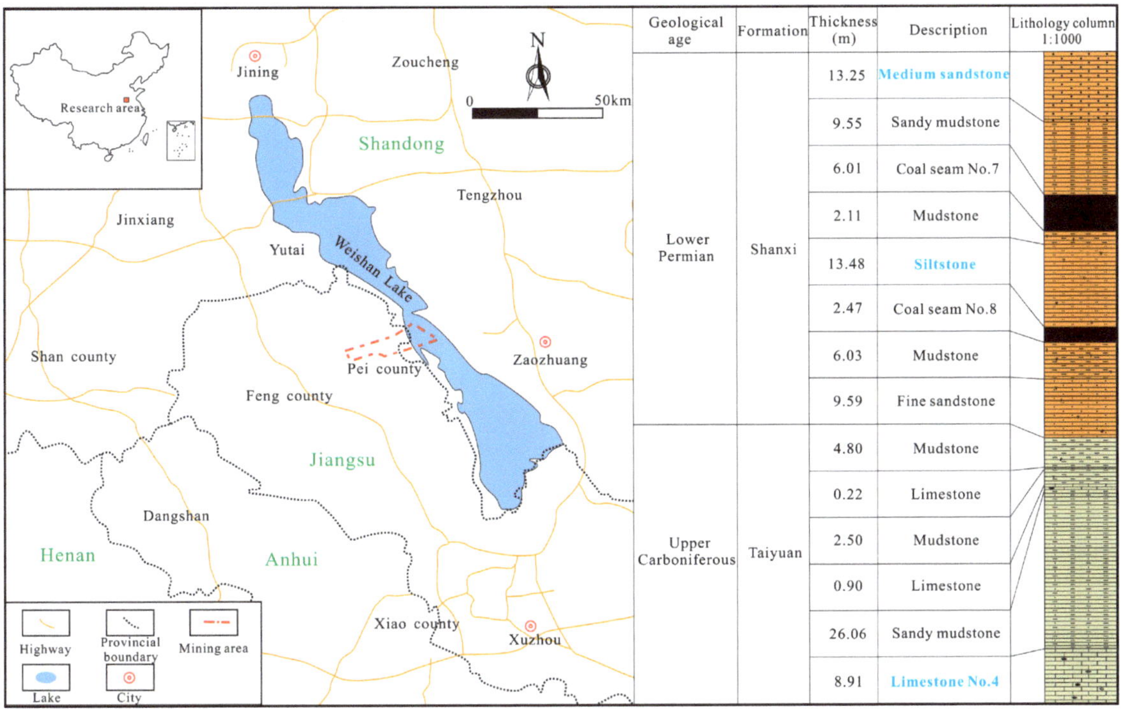

Figure 3. Location and stratigraphic column of coal-bearing strata in the study area.

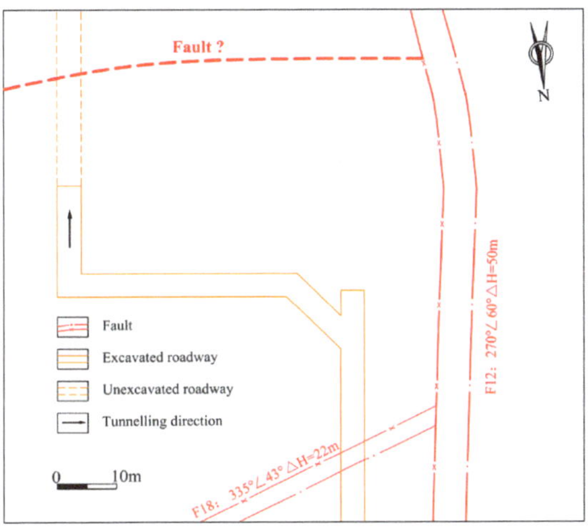

Figure 4. Sketch of structural distribution in the exploration area.

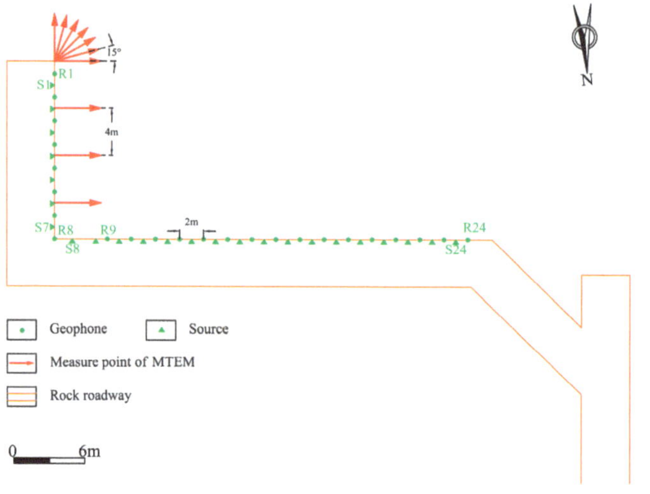

Figure 5. Sketch showing measurement points of the geophysical methods.

After processing the data through static correction, filtering, and automatic gain control, seismic records with nearly zero offset, excited by sources No. 8 to No. 24, were extracted to form a time profile. Compared to other seismic methods, zero-offset seismic exploration reduces the influence of surface and refraction waves. Using a minimal offset helps prevent energy loss from wave mode conversion, allowing more reflection wave energy to be utilized, thereby improving the signal-to-noise ratio for data analysis. As shown in Figure 6a, an oblique reflection interface can be observed to the west of the driving face. Subsequently, all seismic records were used for diffraction stacking migration, which converts time records into spatial distributions. In Figure 6b, the axis of the oblique reflection interface extends from (6, 24) to (26, 43). However, due to the large angle between the reflection interface and the survey line, a discontinuous reflection interface appears in the migration imaging.

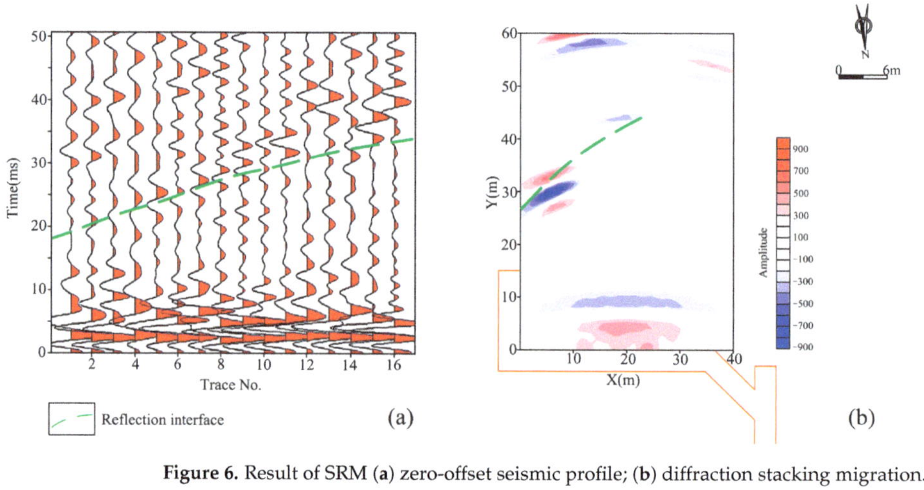

Figure 6. Result of SRM (**a**) zero-offset seismic profile; (**b**) diffraction stacking migration.

(2) Mine transient electromagnetic method

The mine transient electromagnetic method (MTEM) was employed to detect the size and water-bearing properties of the oblique reflection interface. This method is based on

Faraday's law of electromagnetic induction. When current flows through the transmitter coil, the induced primary electromagnetic field propagates into the medium ahead of the coil. Upon switching off the power, good conductors generate a secondary electromagnetic field to counteract the decrease in the primary electromagnetic field. A receiver coil or magnetic probe is used to observe the change in the secondary electromagnetic field over time, allowing hydrogeological information at various depths to be obtained through time-depth conversion. Compared to other geophysical methods, MTEM offers numerous advantages for detecting water-bearing media, such as being non-contact, having good directivity, and high sensitivity to water.

For data acquisition, a square overlapping loop with a side length of 2 m was employed. The measurement points on the right side of the roadway were arranged at intervals of 4 m. To detect the area in front of the heading, the measurement points were increased in density at intervals of 15 degrees at the corner of the driving face. The measurement points and detection direction of the MTEM are shown in Figure 5. Figure 7 presents the contour lines of the apparent resistivity; the blue area represents a structure with lower apparent resistivity. Due to the impact of turn-off time, the MTEM has an exploration blind area of approximately 10 m in the shallow layer. Hence, this region has been subjected to whitening processing. The low apparent resistivity occupies the area ahead of the driving face (Y-axis: 15–32 m) and extends to the right lateral area in front of the roadway. The minimum apparent resistivity is less than 8 Ω·m. Beyond this area, the contour lines of the apparent resistivity gradually increase, indicating that the rocks in these areas are mostly uniform and contain little water. Although this method amplifies the abnormal area due to its volume effect, the low-resistivity zone corresponds well to the oblique reflection interface identified by the seismic method.

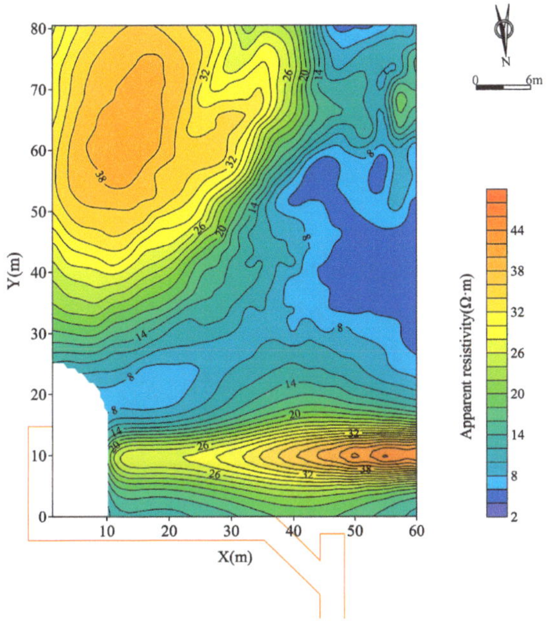

Figure 7. Contour map of apparent resistivity obtained by MTEM.

The results of these two geophysical methods are consistent and complementary, allowing for indirect determination of the location and water-bearing properties of the structures. The hydrogeological parameters provided by the integrated mine geophysical method can inform management measures. However, due to the complex geophysical factors involved, these methods cannot directly identify whether the structure is a fault or another type of geological formation. Therefore, hydrogeological drilling is still necessary to confirm the properties of concealed water-conducting structures and to obtain lithological characteristics and hydrogeological parameters directly.

3.2.2. Hydrogeological Drilling

To identify the properties of the concealed water-conducting structure, three hydrogeological boreholes were drilled to investigate the target region detected by the integrated geological method. Full-section coring was carried out for all boreholes. As shown in Table 1, the No. 4 limestone aquifer was encountered at a depth of 46 m in borehole No. 1. The core recovery rate between 46 and 48 m was low, and the rock core was highly fractured. Meanwhile, groundwater surged into the borehole at a rate of roughly 30 cubic meters per hour and was subsequently pumped out. In borehole No. 2, the No. 4 limestone aquifer was reached at 48.9 m. There was a small initial water inflow from the borehole. The water inflow gradually increased, with a significant rise occurring at 56.5 m, reaching a maximum of approximately 70 cubic meters per hour. In borehole No. 3, the No. 4 limestone aquifer appeared at a depth of 2.5 m. The water inrush peaked at 50 cubic meters per hour between 3.7 and 5.7 m, where numerous fissures were present.

Table 1. The borehole parameters and water inrush.

Borehole No.	Azimuth (°)	Dip (°)	Diameter (mm)	Depth (m)	Maximum Water Inrush (m^3/h)
1	174	−0.5	75	60	30
2	144	12	75	120	70
3	188	17	75	150	50

Through analyzing the core data and stratigraphic succession, the structure detected by the integrated geophysical method was identified as a normal fault with a displacement of approximately 8 m. As shown in Figure 8, there is a slight deviation between the locations detected by drilling and those identified by the integrated geophysical method. Nevertheless, the correlation is satisfactory and improves the location precision compared with using a single method.

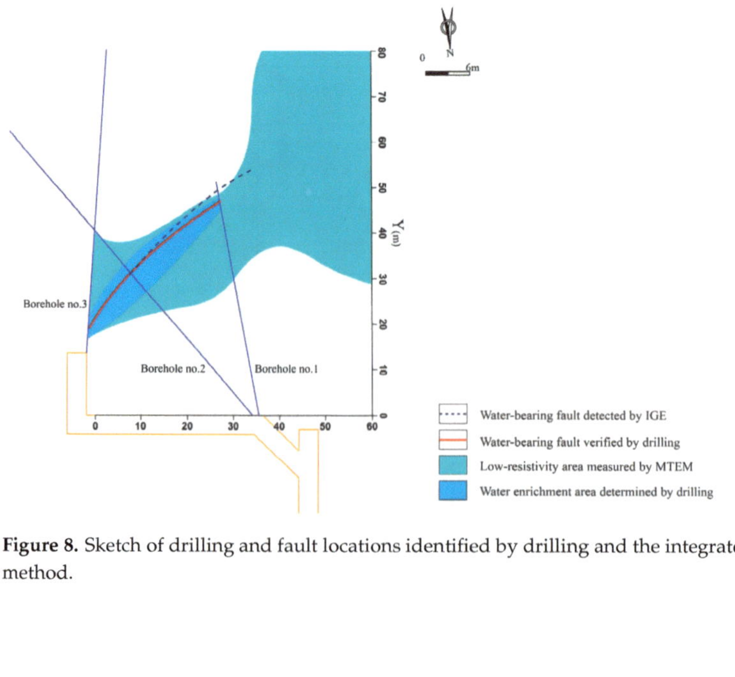

Figure 8. Sketch of drilling and fault locations identified by drilling and the integrated geophysical method.

3.2.3. Hydrochemical Analysis

The inrush water sampled from borehole No. 3 at the driving face was tested to determine its source. The data were compared with the known ionic composition of water samples from the bottom aquifer of the Quaternary system, the sandstone aquifer in the roof of coal seam No. 7, the sandstone aquifer in the roof of coal seam No. 8, the No. 4 limestone aquifer of the Taiyuan Formation, and the Ordovician limestone aquifer. In the known water sources, except for the water sample of the Ordovician limestone aquifer, which was taken from the surface hydrological borehole, the water samples of other aquifers were all actually exposed during excavation work. As the major anions and cations in groundwater, the following seven ions were analyzed: Ca^{2+}, Mg^{2+}, Na^{+}, K^{+}, HCO_3^{-}, Cl^{-}, and SO_4^{2-} [24,25]. The average ion contents of the known water sources and water inrush are presented in Table 2.

Table 2. Ionic composition results in known water sources and inrush water samples.

Water Aquifer	Sample Number	Concentration (mg/L)					
		Ca^{2+}	Mg^{2+}	$Na^{+} + K^{+}$	HCO_3^{-}	Cl^{-}	SO_4^{2-}
Sandstone aquifer in the roof of coal seam No. 7	1	290.20	86.92	216.42	201.97	212.90	1029
	2	306.21	84.03	326.83	211.13	264.80	1223.28
	3	252.90	86.79	268.41	180.61	226.64	1061.52
Sandstone aquifer in the roof of coal seam No. 8	1	372.94	90.32	264.96	200.75	270.45	1280.9
	2	328.05	80.84	229.54	206.85	256.72	1075.31
	3	357.35	92.5	333.73	211.13	271.42	1425.78
Limestone aquifer No. 4	1	88.38	26.62	101.75	319.12	52.34	165.87
	2	25.53	52.28	92.28	295.84	59.72	183.99
	3	23.57	52.28	90.87	270.11	68.74	132.54
	4	47.13	52.28	90.87	270.11	68.74	132.54
Ordovician limestone aquifer	1	335.23	126.44	509.29	237.17	373.25	1674.39
	2	290.66	108.13	604.39	195.32	335.72	1677.68
Sample of water inrush		27.25	38.45	146.97	299.24	74.56	233.79

Using the data in Table 2, a Piper diagram was created to represent water from various aquifers. In the Piper diagram, the distance between two data points reflects the similarity between different water samples. Figure 9 illustrates that the inflow water samples are most similar to the samples from the No. 4 limestone aquifer of the Taiyuan Formation, indicating that the inrush water likely originates from that aquifer. This suggests that there may be a structure conducting water from the No. 4 limestone aquifer into the roadway, as the aquifer is approximately 10 m below the roadway according to the normal formation mapping. Once it is determined that the water inrush originates from the No. 4 limestone layer, taking into account that the connectivity of the No. 4 limestone layer is relatively weak and it is a drainable aquifer, a discharge treatment measure is implemented. In total, approximately 24,000 cubic meters of water is discharged from three boreholes.

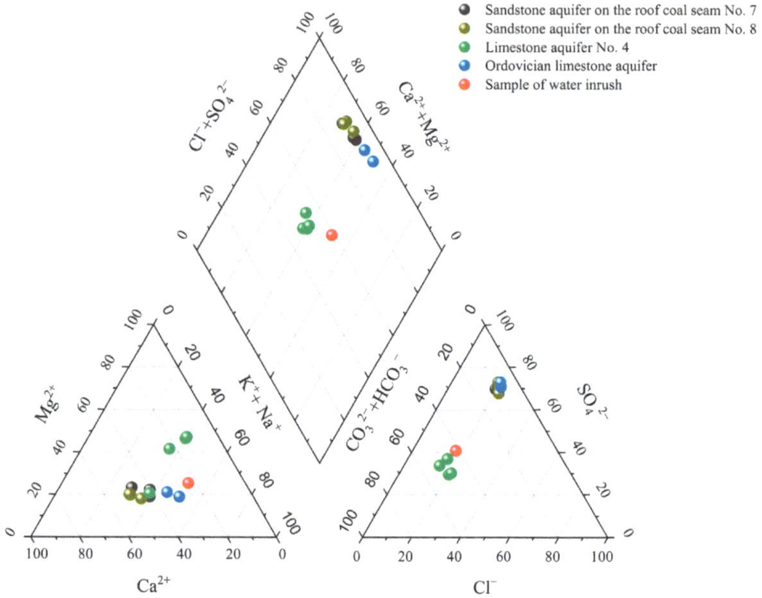

Figure 9. Piper diagram of ionic composition results of known water aquifers and inrush water samples.

4. Discussion

Exploring concealed geological anomalies is critical for effective mine water prevention and control. We propose a coupled approach that integrates geophysical exploration, drilling, and geochemical analysis for the accurate detection of water-conducting structures. This approach follows a systematic process of prediction, detection, verification, and identification. The exploration strategy is structured with geophysical methods leading the way, followed by drilling for verification, and supported by geochemical exploration. Geophysical exploration has lower costs and higher efficiency. It makes drilling construction more targeted. Thus, geophysical exploration is the initial step in water hazard detection. Drilling, which has a higher cost in terms of time and money, is employed to verify the results of geophysical exploration and obtain hydrological parameters. Hydrochemical analysis is used to identify water sources. Its more significant role is to provide a foundation for subsequent treatment measures. By combining these methods, we can mitigate the limitations inherent in using any single technique. This multimethod approach not only improves detection accuracy but also allows for both qualitative and quantitative assessment of geological anomalies.

Despite the effectiveness of this comprehensive exploration system, challenges remain in its application. The first challenge lies in the selection of an appropriate geophysical method, which is crucial as the initial step of exploration. The diverse characteristics of concealed geological structures, coupled with the complexity of the roadway construction environment, complicate this choice. With numerous geophysical methods available, determining the most suitable method is a key factor affecting the detection accuracy. Additionally, while combined detection methods enhance accuracy, they also increase costs and time and can significantly impact coal mine production. Traditional short-distance exploration and excavation techniques are not aligned with the demands of modern intelligent and rapid excavation practices[26]. Therefore, there is a pressing need for new detection methods and technologies that offer faster detection speeds, longer detection ranges, and higher accuracy.

5. Conclusions

This research shows that the accurate detection of concealed water-conducting structures can be achieved by employing a geophysical-drilling-hydrochemical coupled method.

(1) The integrated geophysical method, combining seismic exploration and electrical prospecting, accurately detected the location and water content of concealed geological structures. The results of the geophysical exploration provided a target area for hydrological drilling, thereby making the drilling work more efficient and focused.
(2) Hydrogeological drilling provided rock characteristics and hydrogeological features for both the analysis of the properties of concealed water-conducting structures and the verification of geophysical prospecting results. Meanwhile, it directly exposed the groundwater and obtained water samples for hydrochemical analysis.
(3) Hydrochemical analysis was employed to analyze the inrush water sources and runoff conditions. According to the spatial relations between the source aquifer and mining spaces, the vertical extension range of the concealed water-conducting structures was roughly estimated.

By combining different methods, we can obtain various parameters and information directly or indirectly and analyze the state and hydrogeological characteristics of concealed water-conducting structures from different perspectives. Thus, the limitations and possible bias in the interpretation of a single method or technology can be avoided. This approach can improve the methods for detecting concealed water-conducting structures from the initial location estimation stage through deduction to qualitative exploration and even quantitative detection, which is significant for the safety and production efficiency of coal mining.

Author Contributions: Conceptualization, T.L.; methods, T.L. and B.W.; software, H.L.; validation, H.J.; investigation, T.L. and H.J.; writing—original draft preparation, T.L.; writing—review and editing, H.J.; visualization, B.W.; funding acquisition, T.L. All authors have read and agreed to the published version of the manuscript.

Funding: This research was funded by the National Natural Science Foundation of China, grant number 42304152, and the Jointed Foundation by Shaanxi Science and Technology Plan, grant number 20JK0771.

Data Availability Statement: The datasets generated and/or analyzed during the current study are available from the corresponding author upon reasonable request.

Acknowledgments: Special thanks are extended to the editors and all anonymous reviewers.

Conflicts of Interest: All authors declare that the research was conducted in the absence of any commercial or financial relationships that could be construed as potential conflicts of interest.

References

1. Zeng, Y.; Wu, Q.; Zhao, S.; Miao, Y.; Zhang, Y.; Mei, A.; Meng, S.; Liu, X. Characteristics, causes, and prevention measures of coal mine water hazard accidents in China. *Coal Sci. Technol.* **2023**, *51*, 1–14.
2. Song, Z.; Hao, J.; Shang, J.; Dan, Y. Study on water inrush from fault's prevention and control theory. *J. China Coal Soc.* **2013**, *38*, 1511–1515.
3. Wu, Q. Progress, problems and prospects of prevention and control technology of mine water and reutilization in China. *J. China Coal Soc.* **2014**, *39*, 795–805.
4. Hu, W.; Tian, G. Mine water disaster type and prevention and control countermeasures in China. *Coal Sci. Technol.* **2010**, *38*, 92–96.
5. Xu, Y.; Zhang, E.; Luo, Y.; Yi, K. Mechanism of Water Inrush and Controlling Techniques for Fault-Traversing Roadways with Floor Heave Above Highly Confned Aquifers. *Mine Water Environ.* **2020**, *39*, 320–330. [CrossRef]
6. Gu, H.; Lai, X.; Tao, M.; Momeni, A.; Zhang, Q. Dynamic mechanical mechanism and optimization approach of roadway surrounding coal water infusion for dynamic disaster prevention. *Measurement* **2023**, *223*, 113639. [CrossRef]
7. Zhang, Z.; Gao, Y.; Wu, Q.; Wei, S. Discussion on the technical system of solid prevention and control on mine flooding. *J. China Coal Soc.* **2013**, *38*, 378–383.
8. Chudy, K. Mine Water as Geothermal Resource in Nowa Ruda Region (SW Poland). *Water* **2022**, *14*, 136. [CrossRef]
9. Jin, D.; Zhao, C.; Duan, J.; Qiao, W.; Lu, J.; Li, P.; Zhou, Z.; Li, D. Research on 3D monitoring and intelligent early warning system for water hazard of coal seam floor. *J. Chna Coal Soc.* **2020**, *45*, 2256–2264. [CrossRef]

10. Rey, J.; Martínez, J.; Hidalgo, M.; Mendoza, R.; Sandoval, S. Assessment of Tailings Ponds by a Combination of Electrical (ERT and IP) and Hydrochemical Techniques (Linares, Southern Spain). *Mine Water Environ.* **2021**, *40*, 298–307. [CrossRef]
11. Ding, B. Features and prevention countermeasures of major disasters occurred in China coal mine. *Coal Sci. Technol.* **2017**, *45*, 109–114.
12. Diamond, R.; Staden, C.; Dippenaa, M. Tracing Mine Water Flows in a Dolomite Quarry, South Africa, Using Hydrochemistry and Stable Isotopes. *Mine Water Environ.* **2024**, *43*, 278–293. [CrossRef]
13. Han, D.; Li, D.; Cheng, J. DC method of advanced detecting disastrous water-conducting or water-bearing geological structures along same layer. *J. China Coal Soc.* **2010**, *35*, 635–639.
14. Teixeira, F.; Chew, W. Finite-Difference computation of transient electromagnetic waves for cylindrical Geometries in complex media. *IEEE Trans. Geosci. Remote Sens.* **2000**, *38*, 1530–1543. [CrossRef]
15. Yu, J.; Liu, Z.; Liu, S. Theoretical analysis of mine transient electromagnetic method and its application in detecting water burst structures in deep coal stope. *J. China Coal Soc.* **2007**, *32*, 818–821.
16. Danielson, J.; Auken, E.; Jorgensen, F. The application of the transient electromagnetic method in hydro-geophysical surveys. *J. Appl. Geophys.* **2003**, *53*, 181–198. [CrossRef]
17. Bohlen, T.; Borm, G.; Giese, R.; Klose, C.; Mielitz, S.; Otto, P. ISIS-integrated seismic imaging system for the geological prediction ahead of underground construction. *Eur. Assoc. Geosci. Eng.* **2003**, cp-6-00209. [CrossRef]
18. Otto, R.; Button, E.; Bretterebner, H.; Schwab, P. The application of TRT-True Reflection Tomography-at the Unterwald tunnel. *Felsbau* **2002**, *20*, 51–56.
19. Sattel, G.; Frey, P.; Amberg, R. Prediction ahead of the tunnel face by seismic methods-pilot project in Centovalli Tunnel, Locarno, Switzerland. *First Break* **1992**, *10*, 19–25. [CrossRef]
20. Liu, S.; Liu, J.; Yue, J. Development status and key problems of Chinese mining geophysical technology. *J. China Coal Soc.* **2014**, *39*, 19–25.
21. Lu, T.; Liu, S.; Wang, B.; Wu, R.; Hu, X. A Review of Geophysical Exploration Technology for Mine Water Disaster in China: Applications and Trends. *Mine Water Environ.* **2017**, *36*, 331–340. [CrossRef]
22. Wang, Q.; Yang, T. The seismic reflection method predicting tunnel geological hazards. *Prog. Geophys.* **2006**, *21*, 643–649.
23. Dong, D.; Zhang, J. Discrimination Methods of Mine Inrush Water Source. *Water* **2023**, *15*, 3237. [CrossRef]
24. Li, Y.; Sun, Y.; Xu, Z.; Zhang, G.; Liu, Y.; Ding, J. Analysis of composition of mine inflow from complicated multi-aquifer affecting safety production in coal mine. *J. Min. Saf. Eng.* **2010**, *3*, 433–437.
25. Wei, W.; Lu, X.; Shi, L. Identification method of multi-water source of mine water inrush. *J. China Coal Soc.* **2010**, *5*, 811–815.
26. Yin, S.; Xu, B.; Yin, H.; Cao, M.; Ding, Y.; Liang, M. Basic structure and connotation of mine water prevention and control discipline. *Coal Sci. Technol.* **2023**, *51*, 24–35.

water

MDPI

Article

Intrusion Event Classification of a Drainage Tunnel Based on Principal Component Analysis and Neural Networking

Peng Yuan [1], Weihao Zhang [1], Xueyi Shang [2,*] and Yuanyuan Pu [2]

[1] China Yangtze Power Co., Ltd., Yibin 644612, China; yuan_peng@ctg.com.cn (P.Y.); zhang_weihao@ctg.com.cn (W.Z.)
[2] School of Resources and Safety Engineering, Chongqing University, Chongqing 400044, China; yuanyuanpu@cqu.edu.cn
* Correspondence: shangxueyi@cqu.edu.cn

Abstract: Drainage tunnel stability is crucial for engineering project safety (e.g., mine engineering and dams), and rockfall events and water release are key indicators of drainage tunnel stability. To address this, we developed a monitoring system to simulate drainage tunnel intrusions based on distributed acoustic sensing (DAS), and we obtained typical characteristics of events like rockfall events and water release. Given the multitude of DAS signal feature parameters and challenges, such as high-dimensional features impacting the classification accuracy of machine learning, we proposed an identification method for drainage tunnel intrusion events using principal component analysis (PCA) and neural networks. PCA reveals that amplitude-related parameters—amplitude, mean amplitude, and energy—significantly contribute to DAS signal classification, reducing the feature parameter dimensions by 54.8%. The accuracy of intrusion event classification improves with PCA-processed data compared to unprocessed data, with overall accuracy rates of 79.1% for rockfall events and 72.7% for water release events. Additionally, the artificial neural network model outperforms the Bayesian and logistic regression models, demonstrating that ANN has advantages in handling complex models for intrusion event classification.

Keywords: drainage tunnel; intrusion event; classification model; principal component analysis; neural network

Citation: Yuan, P.; Zhang, W.; Shang, X.; Pu, Y. Intrusion Event Classification of a Drainage Tunnel Based on Principal Component Analysis and Neural Networking. *Water* **2024**, *16*, 2409. https://doi.org/10.3390/w16172409

Academic Editors: Christos S. Akratos and Cesar Andrade

Received: 18 July 2024
Revised: 17 August 2024
Accepted: 21 August 2024
Published: 27 August 2024

1. Introduction

The drainage tunnel stability is crucial for the safety of engineering projects (e.g., mine engineering and dams). However, it can be affected by rock lithology, rock mass fault, seepage water, etc. Drainage tunnel stability monitoring is getting more and more attention [1–3]. Common engineering monitoring methods include rock mass stress and displacement monitoring, satellite remote sensing, 3D laser scanning, geological radar, microseismic monitoring, and distributed acoustic sensing (DAS). Rock mass stress and displacement monitoring is simple but limited to point monitoring and cannot reflect intrusion events like personnel inspections. Satellite remote sensing can only monitor surface displacement [4] and cannot observe internal damage or locate intrusion events within drainage tunnels. Three-dimensional laser scanning can capture the shape of drainage tunnels but cannot penetrate the internal structure of rock mass or detect intrusion events like personnel inspections or rock mass collapses [5]. Microseismic monitoring technology is well suited for monitoring rock mass fractures but struggles to detect small-scale intrusion events, and it cannot provide real-time locations for personnel inspections or animal intrusions [6]. DAS technology, with its high sensitivity and stability [7,8], is highly suitable for monitoring various intrusion events, where type classification of intrusion events is crucial in DAS monitoring for drainage tunnels.

Initially, intrusion event classification in DAS monitoring was based on manual observation of signals and heavily influenced by subjective judgment [9]. To handle this,

Fast Fourier Transform (FFT) was adopted to convert DAS signals from time to frequency domain [10], which provides a fresh perspective for classifying intrusion events. However, due to factors such as signal propagation distance, frequency bands of different intrusion types often overlap and reduce the classification accuracy of the frequency-based methods. Furthermore, parameters like amplitude threshold, short-term energy, amplitude zero-crossing rate, and Mel Frequency Cepstral Coefficients were employed for intrusion event classification [11,12]. Additionally, techniques such as wavelet decomposition and empirical mode decomposition were introduced to perform multi-scale extraction of DAS signal features [13,14], thereby enriching the DAS signal feature database. Based on these characteristics, machine learning methods such as Support Vector Machines, Random Forests, Artificial Neural Networks, and Hidden Markov Models have also been applied to classify DAS signals [15–17].

In recent years, time–frequency techniques have been employed to classify intrusion events. For example, Xu et al. [18] used the Short-Time Fourier Transform (STFT) to transform DAS signals into the time–frequency domain, which then served as input for a Convolutional Neural Network (CNN), achieving notable success in classifying intrusion events. Furthermore, Li et al. [19] utilized CNN to extract features from multi-channel signals in spatial DAS systems. They used Long Short-Term Memory (LSTM) to analyze temporal relationships, achieving real-time monitoring of DAS signal intrusions, although with a low detection rate and slow response time. In conclusion, the time domain features of DAS signals are still commonly used to classify intrusion events. However, the extraction of numerous feature parameters can limit the classification accuracy of machine learning, and it is unclear which parameters are crucial for the precise classification of DAS signals.

Given the numerous feature parameters of DAS signals, which may reduce the accuracy of machine learning-based classification, and the absence of an established method for classifying intrusion events in drainage tunnel environments, this study proposed a model based on principal component analysis (PCA) and neural networks for intrusion event classification. Firstly, laboratory experiments adapted to the drainage tunnel environment were conducted to capture DAS signals from intrusion events. Subsequently, these DAS signals and their multi-scale features were extracted to build a feature library. Then, the PCA was used to evaluate the importance of each feature parameter, facilitating data dimensionality reduction. Finally, the post-dimensionality reduction data were served as the input of the neural network. By comparing the effectiveness of intrusion event classification before and after PCA dimensionality reduction, this study aims to provide a new and effective method for classifying intrusion events in drainage tunnel environments.

2. Methodology

2.1. Parameter Selection and Calculation Principles

As stated in the Introduction section, it is known that DAS signals have an extensive number of feature parameters. This study focused on commonly used features for intrusion event classification. The selected features include signal duration T, maximum amplitude A, mean amplitude μ, variance σ_2, rise time RT, ring count N, energy E, Kurtosis K, Skewness S, dominant frequency f_{dom}, median frequency f_{med}, fractal dimension L, signal entropy H, wavelet energy ratio P_1 to P_6, wavelet fractal dimension L_1 to L_6, and wavelet signal entropy H_1 to H_6. Typical DAS signal feature parameters are shown in Figure 1. Then, the parameters' computation from T to f_{med} can be found in Grosse et al. [20], the fractal dimension is calculated using the Higuchi algorithm [21], and the computation methods of signal entropy and wavelet-coefficient-related parameters are given as follows:

(1) Signal Entropy

Signal entropy is often used to measure the uncertainty of a signal or probability distribution. Assume that the DAS signal time series is $x(i)$ (i = 1, 2, . . ., N) and divide the $x(i)$ into j intervals (buckets, j = 1, 2, . . ., M) according to the signal amplitudes. Then, divide the number of data points N_j in each bucket by the total number of points N; the probability

$P_j = N_j/N$ for each bucket is obtained, from which the signal entropy is calculated by the following:

$$H = -\sum_{i=1}^{M} P_j \log_2 P_j \quad (1)$$

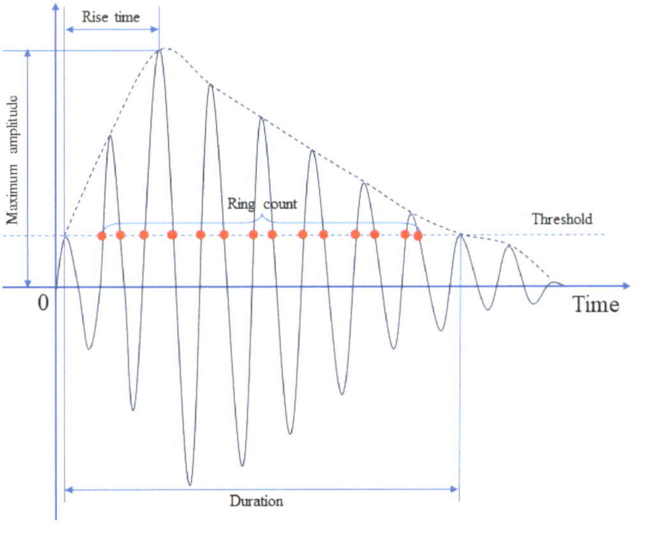

Figure 1. Schematic diagram of calculation principle of typical signal characteristics.

(2) Wavelet-Coefficient-Related Parameters

The wavelet basis db3 was selected to decompose the DAS signal into five levels, resulting in five approximation coefficients and one detail coefficient, denoted as C_1~C_6. The fractal dimension and signal entropy calculation methods based on wavelet coefficients are similar to those mentioned above, requiring only the replacement of the signal with wavelet coefficients. The process to calculate the energy ratio of wavelet coefficients is given as follows: the energy of each wavelet coefficient is represented by $E_k = \sum_{k=1}^{6} C_k^2, k = 1, 2, \ldots, 6$, and the energy ratio of each coefficient is calculated as:

$$P_k = E_k / \sum_{k=1}^{6} E_k \quad (2)$$

2.2. Principal Component Analysis

The feature parameters selected in Section 2.1 are used to classify DAS signals. However, employing these 31 features in a neural network or similar classifiers can reduce the classification accuracy. To address this, PCA is adopted to reduce the dimensionality of the feature parameter matrix while retaining essential information as much as possible. This method identifies correlations within the data and extracts key features (i.e., principal components), transforming them into a new coordinate system where each axis represents a primary variable direction of the original data. The main steps of PCA for the DAS signal feature parameter matrix are given as follows:

① Data normalization: Each feature is normalized to ensure feature comparability. Feature normalization is achieved by subtracting the mean of each feature and dividing by its standard deviation, as follows:

$$Z_{ij} = \left(X_{ij} - \overline{X}_j\right)/\sigma_j \tag{3}$$

where X_{ij} is the jth feature value of the ith signal, $\overline{X}_j$ the mean value, and σ_j the standard deviation.

② Covariance matrix construction: Then, the covariance matrix of the normalized data Z_{ij} is calculated. The covariance matrix elements are the covariances between features, reflecting their interrelationships. The formula for calculating the covariance matrix is as follows:

$$\text{Cov}(Z_i, Z_j) = \frac{1}{n-1}\sum_{k=1}^{n} Z_{ki}Z_{kj} \tag{4}$$

where Z_{ki} and Z_{ji} are the ith and jth feature values of the kth signal, respectively.

③ Eigenvalue and eigenvector calculation: The covariance matrix is decomposed to find its eigenvalues and corresponding eigenvectors. This step reveals the data primary directions, where the magnitude of an eigenvalue indicates the variance (i.e., the amount of information) in that direction, and the eigenvectors define these directions. The formula for eigendecomposition is as follows:

$$\mathbf{Cv} = \lambda\mathbf{v} \tag{5}$$

where $\mathbf{C}$ is the covariance matrix, λ represents an eigenvalue, and $\mathbf{v}$ is the corresponding eigenvector.

④ Selection of principal components and data transformation: The most significant eigenvectors are selected based on the magnitude of the eigenvalues (usually those associated with the largest values that account for 80% of the sum of the eigenvalues [22]). These eigenvectors form the basis of the new feature space. The original data are then projected onto this basis to create a new reduced dataset. The data transformation formula is as follows:

$$\mathbf{T} = \mathbf{ZP} \tag{6}$$

where $\mathbf{Z}$ is the normalized data matrix, $\mathbf{P}$ is the matrix composed of selected eigenvectors, and $\mathbf{T}$ is the transformed data matrix.

Through the above principal component analysis, we can obtain insights into the primary variables and their interactions in the data and reduce the data complexity, thereby potentially enhancing the accuracy of DAS signal classification.

2.3. PCA-ANN Classification Model

The feature vector $\mathbf{T}$ obtained from the principal component analysis is used to construct the PCA-ANN model for DAS signal classification, as shown in Figure 2. The input layer has 14 neurons, corresponding to the principal component number extracted by PCA (Section 4.1), and there are 29 neurons in the hidden layer according to 2 × neurons + 1 of the input layer [23]; the output layer is designated for DAS signal classification. The weight matrix from the input layer to the hidden layer is denoted as $\mathbf{W}^{(1)}$ (size 14 × 29), and the bias vector is $\mathbf{b}^{(1)}$ (length 29). The weight matrix from the hidden layer to the output layer is $\mathbf{W}^{(2)}$ (size 29 × 1), with a bias $b^{(2)}$ (single value).

The transfer formula from the input layer to the hidden layer is:

$$\mathbf{h} = f(\mathbf{W}^{(1)}\mathbf{T} + \mathbf{b}^{(1)}) \tag{7}$$

where f is the activation function for the hidden layer, chosen as the Sigmoid activa tion function.

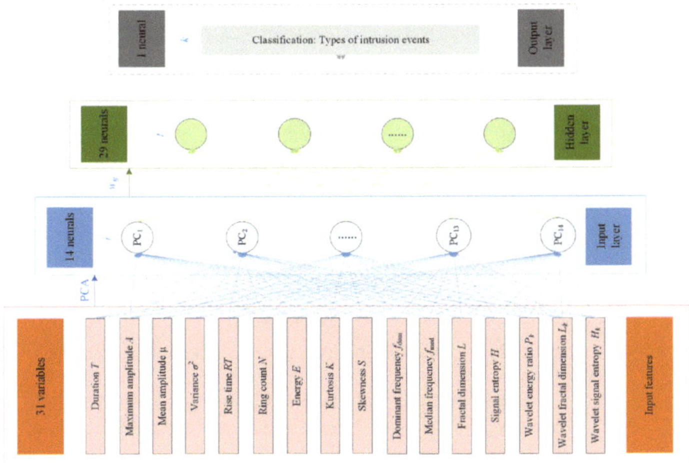

Figure 2. PCA-ANN-based DAS signal classification model.

The transfer formula from the hidden layer to the output layer is as follows:

$$y = g(\mathbf{W}^{(2)}\mathbf{h} + b^{(2)}) \tag{8}$$

where g is the activation function for the hidden layer, chosen as the softmax activation function.

3. Intrusion Event Simulation

3.1. Experiment Setup

This study focuses on addressing issues threatening the safety of a hydropower station's slopes, personnel, and equipment. These issues include roof collapses in drainage tunnels, water release, and intrusions by outsiders and animals. To handle this, a DAS monitoring system was set up in Area A of the air raid shelter at Chongqing University (experimental details and monitoring principles as shown in Figure 3) to simulate the environment of the drainage tunnel at the hydropower station and intrusion events. Based on this, research on classifying intrusion events was conducted to support the safety monitoring of the drainage tunnel.

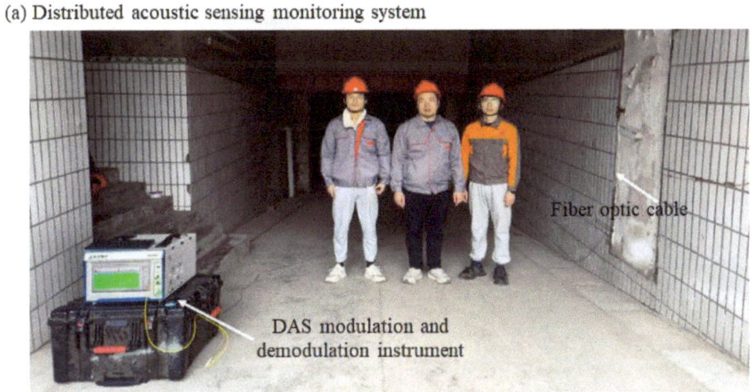

Figure 3. *Cont.*

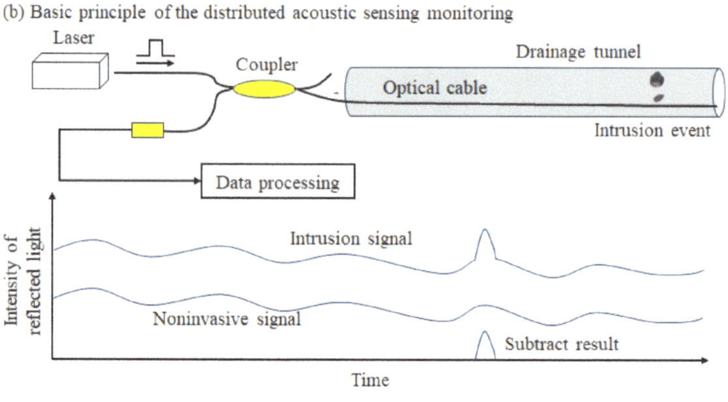

Figure 3. DAS monitoring experimental system and its principle.

The detailed experiment (Figure 4) is conducted as follows: ① Personnel inspection: Three scenarios were designed, involving one, two, and three personnel, respectively. Inspection personnel will move at the tunnel entrance and distances of 60 m, 120 m, and 300 m, with a walking speed of 0.4 m/s and a running speed of 3 m/s, while recording the corresponding acoustic signals. ② Animal intrusion simulation: Sounds of snakes crawling, bats flying, mice moving, and wild boars walking and howling were downloaded from an internet sound library. Using speakers, these sounds are played at 40 dB, 60 dB, and 80 dB. Additionally, a toy snake is dragged to simulate snake intrusions. ③ Rockall simulation: Rocks weighing 1 kg, 3 kg, and 5 kg are dropped from heights of 0.5 m, 1.5 m, and 3 m to simulate rock collapse intrusion events. ④ Water release simulation: Water flows of 0.5 L/s, 1 L/s, and 2 L/s are dropped from heights of 0.5 m, 1 m, and 2 m to simulate seepage water dropping intrusion events, where the water release rates are controlled by a valve. ⑤ Vehicle intrusion simulation: Driving past the air raid shelter to simulate a vehicle passing by the drainage tunnel. During the experiment, the base parameters for the demodulator are as follows: Sampling frequency, start distance, and stop distance are set to 100 Hz, 2.76 m, and 53.71 m, respectively. System settings include acquisition mode in continuous mode, fiber type as single-mode fiber, and refractive index at 1.468.

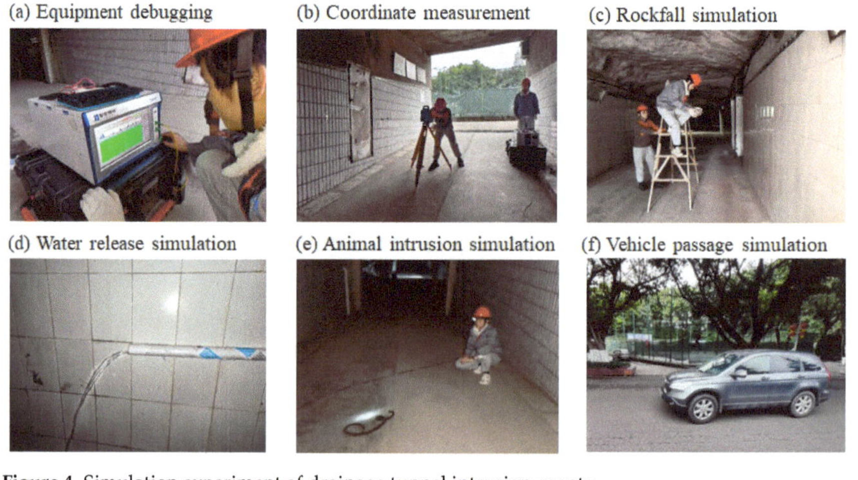

Figure 4. Simulation experiment of drainage tunnel intrusion events.

3.2. Character Analysis of Intrusion Event Signals

The experiment obtained typical acoustic signals of rockfall events, water release, personnel inspection, animal intrusion, vehicle passage, and background noise are shown in Figure 5. It is evident that the vehicle passage signal has a low signal-to-noise ratio due to the distance from the car to the sensor. The personnel inspection signal also has a low signal-to-noise ratio; it fluctuates as the amplitude increases when personnel approach the sensor and decreases as they move away. The snake crawling event displays distinct repetitiveness due to multiple simulations of snake movement. The rockfall event and water release show pronounced initial vibrations resulting from the rapid impact of rockfalls or water release.

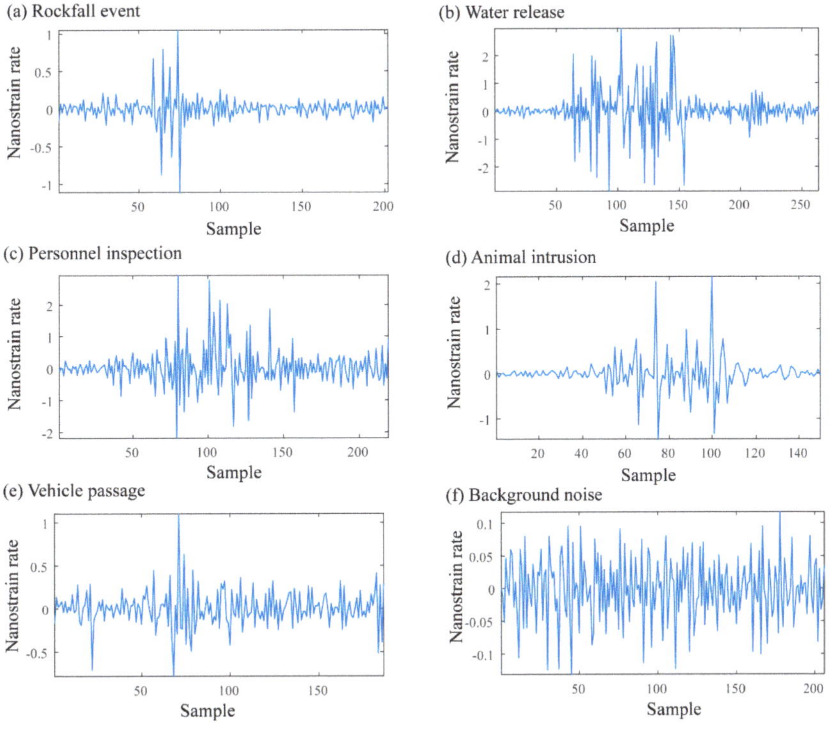

Figure 5. Typical DAS signals of intrusion events.

Then, the computation theories shown in Section 2.1 were used to obtain the acoustic signal characteristics of Figure 5, and the 31 characteristic values are listed in Table 1. There are some differences in the acoustic signal parameters among different types of intrusion events, suggesting a need to understand the statistical characteristics of these parameters more generally. The box plots of typical features for each type of intrusion signal are displayed in Figure 6, with the background noise selected from signals being automatically detected from continuous background signals. These box plots illustrate the disparities among features, with water release signals showing relatively significant differences from other intrusion event types and some features of background noise (mean, rise time, energy, etc.) also differing markedly from other intrusion event types. The feature extraction results indicate that, due to the signal complexity of drainage tunnel intrusion events, it is challenging to classify events using single or simple feature parameters accurately. Thus, there is a need to develop methods based on multi-parameter or image deep learning to classify intrusion events accurately.

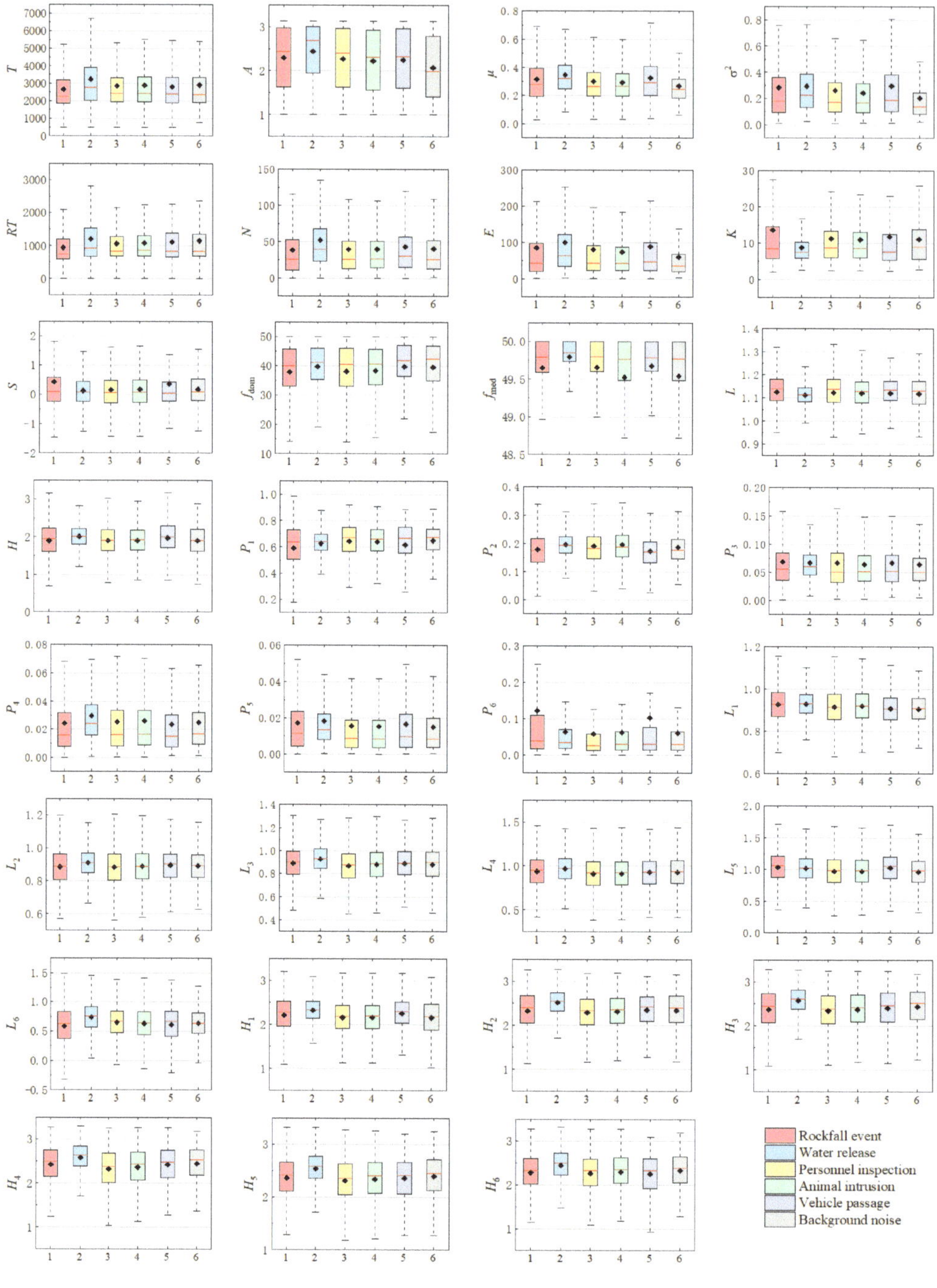

Figure 6. Box plots of typical features for each type of intrusion event. The numbers 1 to 6 on the horizontal axis of each box plot correspond to the rockfall event, water release, personnel inspection, animal intrusion, vehicle passage, and background noise, respectively.

Table 1. Acoustic signal characteristics for the events shown in Figure 5.

No.	Parameter	Vehicle Passage	Personnel Inspection	Animal Intrusion	Rockfall Event	Water Release	Background Noise
1	T	1850.00	2180.00	1490.00	2010.00	2630.00	/
2	A	1.10	2.92	2.16	1.12	2.98	0.13
3	μ	0.16	0.39	0.21	0.10	0.42	0.04
4	σ^2	0.05	0.37	0.16	0.03	0.55	0.00
5	RT	700.00	790.00	990.00	730.00	1020.00	/
6	N	26.00	20.00	10.00	7.00	23.00	/
7	E	8.75	81.22	23.62	6.74	144.88	0.49
8	K	6.92	8.62	14.07	19.07	8.32	2.84
9	S	0.36	1.00	1.32	−0.21	0.02	−0.26
10	f_{dom}	39.78	47.49	42.67	34.65	40.53	40.78
11	f_{med}	49.46	49.77	50.00	50.00	49.62	50.00
12	L	1.17	1.15	1.05	1.17	1.09	1.16
13	H	2.14	2.04	1.84	1.39	2.02	2.95
14	P_1	0.71	0.66	0.67	0.84	0.65	0.69
15	P_2	0.16	0.15	0.28	0.10	0.16	0.20
16	P_3	0.09	0.05	0.04	0.03	0.11	0.06
17	P_4	0.01	0.06	0.00	0.01	0.06	0.01
18	P_5	0.01	0.01	0.00	0.00	0.00	0.00
19	P_6	0.02	0.07	0.00	0.02	0.01	0.03
20	L_1	0.73	0.87	0.70	0.91	0.91	0.95
21	L_2	0.95	0.79	0.88	0.90	1.09	0.77
22	L_3	0.80	0.98	0.55	1.06	0.88	1.08
23	L_4	0.47	1.00	0.25	0.70	0.83	0.77
24	L_5	1.12	1.16	0.42	1.53	0.70	1.44
25	L_6	0.86	0.93	0.51	0.27	0.56	0.45
26	H_1	2.48	2.10	1.92	1.32	1.97	3.00
27	H_2	2.27	2.19	1.96	1.78	2.28	2.89
28	H_3	2.95	2.45	2.43	2.62	1.84	2.77
29	H_4	3.02	2.44	2.30	2.70	2.12	2.82
30	H_5	2.92	2.48	2.28	2.37	2.50	2.66
31	H_6	1.57	1.97	2.11	2.41	2.14	2.37

Note: The background noise data were randomly extracted from periods without any intrusion experiments. Many of these signals, similar to Figure 5f, have no significant amplitude anomalies, making it impossible to define parameters such as signal duration (T), rise time (RT), and ring count (N).

4. Result and Discussion

4.1. PCA Analysis

The correlation matrix for the parameters calculated by Equation (4) is shown in Figure 7. It can be seen that there are relatively significant correlations between some parameters. Specifically, a strong correlation exists between the DAS signal amplitude, mean, and variance, suggesting that signal amplitude and dispersion are interrelated. The signal duration significantly correlates with rise time and ring count, which is attributed to the duration encompassing rise time and the number of threshold crossings within the signal duration. Skewness exhibits negative correlations with multiple variables, while signal entropy shows positive correlations. Additionally, wavelet coefficient energy ratios generally display negative correlations, whereas wavelet signal entropies show positive correlations. This complexity in correlations among the feature parameters highlights the necessity of conducting PCA to reduce the impact of these correlations.

Then, Equation (5) was used to obtain the eigenvalues of the correlation matrix depicted in Figure 7. These eigenvalues are arranged in descending order, and their cumulative contribution is calculated, as shown in Figure 8. It is apparent that the first four principal eigenvalues decrease rapidly, accounting for 48.71% of the total variance. The decline in eigenvalues slows after that, with the 14th eigenvalue reaching the 80% cumulative contribution threshold. Therefore, the first 14 eigenvectors corresponding to these eigenvalues are used with Equation (6) to reconstruct the feature parameter matrix.

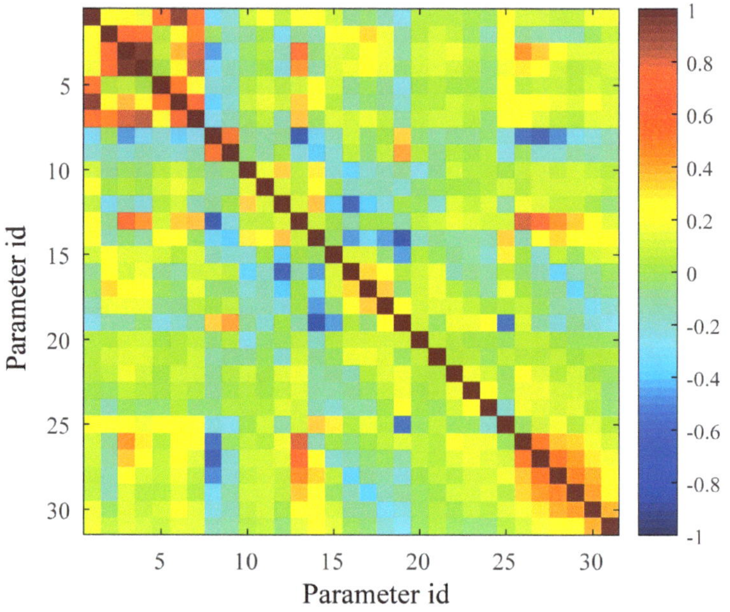

Figure 7. Correlation matrix of the 31 parameters. The parameter id corresponds to that shown in Table 1.

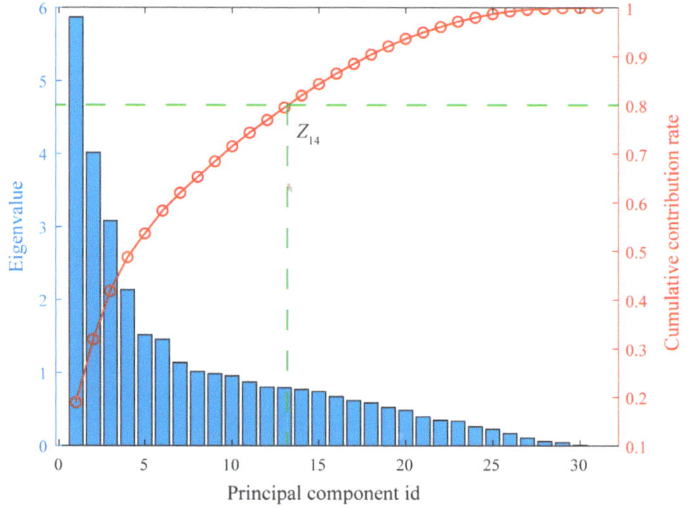

Figure 8. Eigenvalues and cumulative contribution of principal components.

Furthermore, the component loadings of the first four principal components are depicted in Figure 9. This visualization indicates that signal duration, mean, variance, rise time, ring count, energy, Kurtosis, Skewness, signal entropy, and wavelet signal entropy contribute significantly to PC_1. Signal maximum amplitude, mean, variance, energy, dominant frequency, fractal dimension, and wavelet energy ratios contribute significantly to PC_2. Overall, parameters related to signal amplitude, such as amplitude, mean, and energy, make substantial contributions to DAS signal classification, followed by signal entropy.

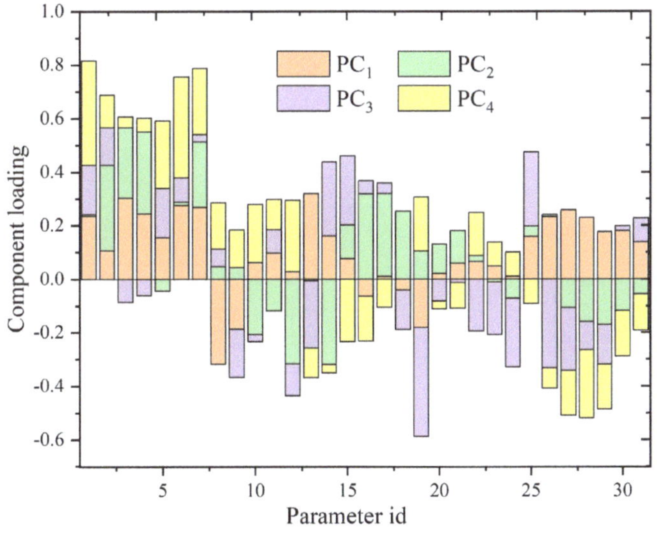

Figure 9. Component loadings of the first four principal components.

4.2. ANN Classification

In Section 4.1, the feature parameter matrix reconstructed after PCA is used as the input for the ANN to conduct intrusion event classification. Given the emphasis on rockfall and water release events in drainage tunnels, this study showcases the classification results for these two types of intrusion events. For simplification, the rockfall event and water release are marked as "1" and "2", respectively. Moreover, the results from the ANN using the normalized original feature parameter matrix are also displayed. The proportions of data for training, validation, and test in the ANN are approximately 70%, 15%, and 15%, respectively. As shown in Figure 10, the loss curves of training, validation, test, and all datasets demonstrate good convergence, indicating that the ANN model possesses strong generalization capabilities. Furthermore, classifications made with the PCA-processed data show smaller average errors for DAS signal classes, confirming the necessity of PCA processing.

Moreover, the training, validation, test, and all dataset-based ROC curves for the ANN are presented in Figure 11. The larger the area under the ROC curve on the lower right side, the better the classification performance. It is observed that the training, validation, test, and all datasets-based ROC curves exhibit the following similar patterns: the True Positive Rate (TPR) for rockfall events is better than that for water release when the False Positive Rate (FPR) is low; however, the TPR for rockfall events is worse than that for water release when the FPR is high. For the same type of event, the ROC curves processed with PCA are positioned above those not processed with PCA, indicating that the PCA-processed model offers better classification results.

The classification accuracy of rockfall and water release events is shown in Figure 12, suggesting that the intrusion event classification performance is similar across training, validation, and test datasets, demonstrating the practical fitness of the constructed ANN model. The overall higher classification accuracy in the training, validation, and test data processed with PCA underscores the effectiveness of PCA processing. Specifically, the classification accuracies for intrusion events based on data before and after PCA processing reached 73.6% and 75.6%, respectively, with the PCA-processed data achieving overall classification accuracies of 79.1% for rockfall events and 72.7% for water release.

The above analysis shows that the intrusion event classification performance of the ANN with PCA-processed data surpasses that of the data without PCA processing. This improvement is primarily due to the principal components obtained through PCA retaining the primary variance information of the data, which reduces noise in the data and highlights essential features [24], thereby reducing the complexity and potential overfitting of the ANN model and enhancing the accuracy of the DAS signal classification.

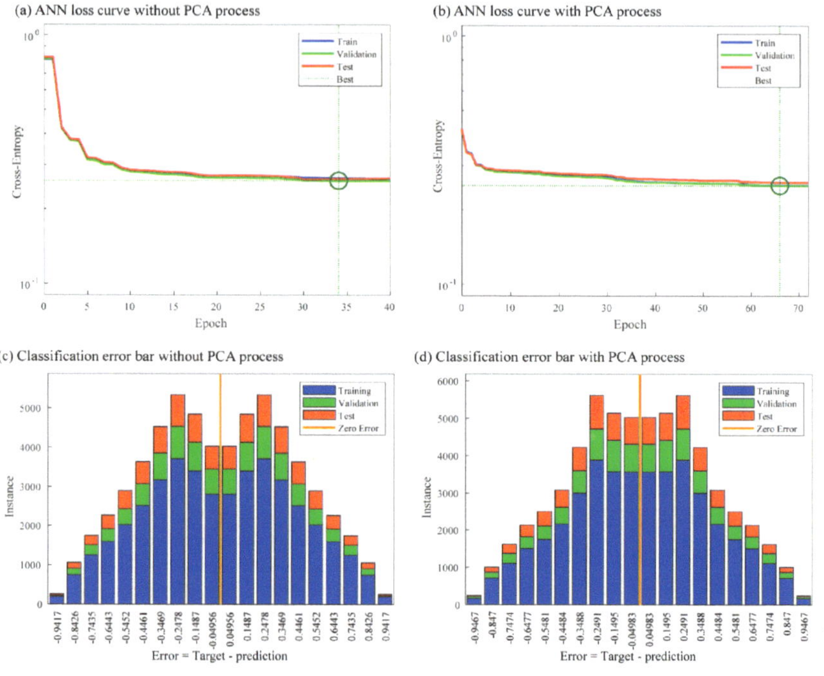

Figure 10. ANN loss curve and classification error before and after the PCA process.

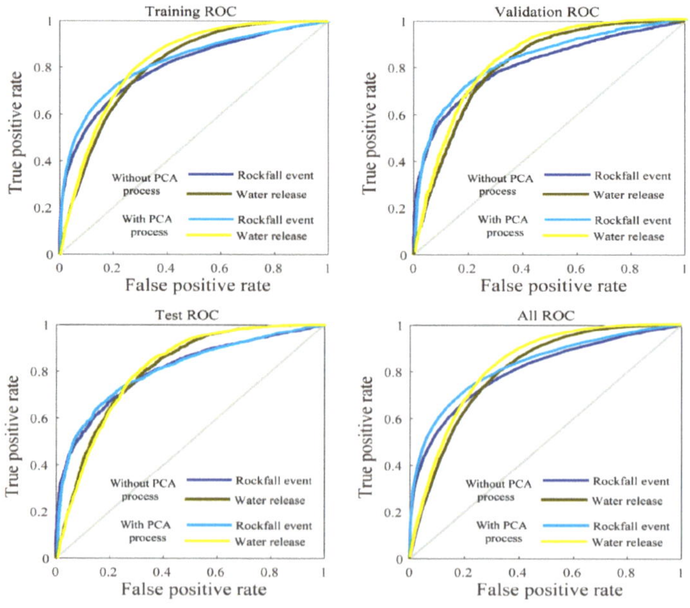

Figure 11. ROC curves before and after PCA processing.

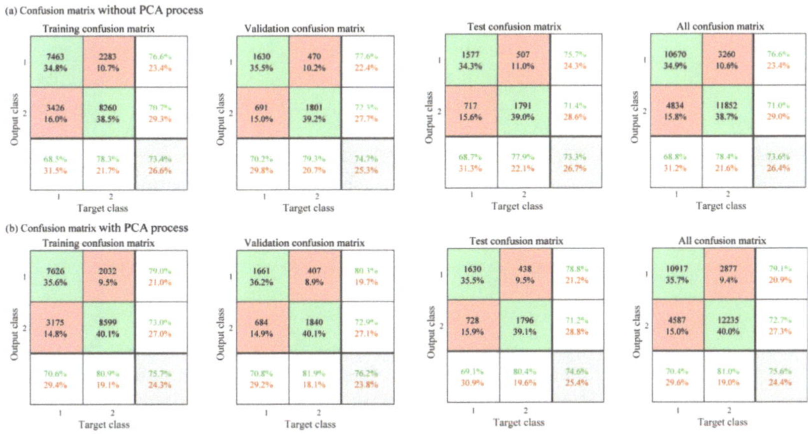

Figure 12. DAS signal classification accuracy before and after PCA processing.

4.3. Comparison with Other Classification Models

Further analysis involves comparing the commonly used Bayesian and logistic regression models (Figure 13). It is evident that both the Bayesian and logistic regression models exhibit lower classification accuracies than the ANN model described in this study, with the logistic regression model slightly outperforming the Bayesian model. The logistic regression model's overall classification accuracies for rockfall events and water release are 66.1% and 74.3%, respectively. The superior performance of the ANN is attributed to its use of multiple layers and nonlinear activation functions, which enable it to learn and model complex patterns and nonlinear relationships in the data [25]. Consequently, ANNs generally perform better in handling complex classification tasks. The performance of the Bayesian model partially depends on its feature independence assumption [26]. Despite being a linear model, logistic regression can be highly effective if the data can be linearly separated after transformations such as PCA. Then, the logistic regression models generally have fewer parameters to estimate compared to the Bayesian model. The reasons discussed above explain why, in this model, ANNs achieve higher classification accuracy compared to Bayesian and logistic regression models and why logistic regression models outperform Bayesian models.

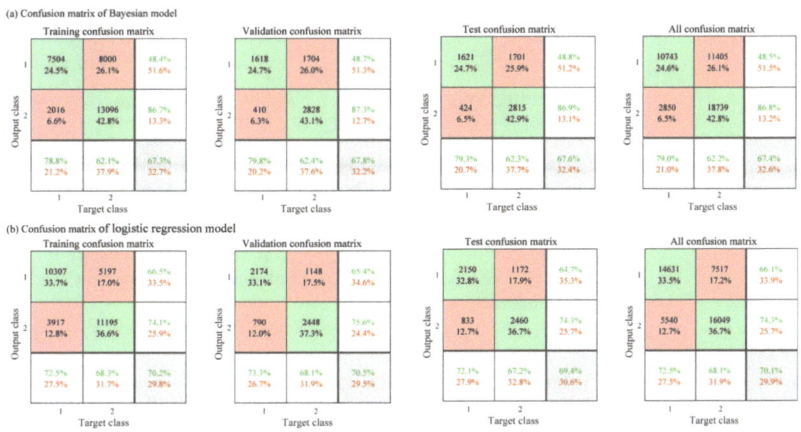

Figure 13. Comparison of classification accuracy using PCA-processed data: Bayesian model vs. logistic regression.

4.4. How to Further Improve the Intrusion Event Classification Accuracy

The above comparisons show the superiority of our method; however, the overall recognition accuracy of 75.6% is relatively low. This is due to our fiber optics being suspended rather than closely attached to the wall in the experiments, which reduces the signal quality. Additionally, we included not only the most characteristic signals from each intrusion event but also signals from adjacent channels, some of which were of lower quality. Therefore, we can optimize the setup of the fiber optics and select higher quality signals to improve classification effectiveness. Further, we can explore more signal feature parameters and other machine learning classification methods or employ deep learning based on image recognition to enhance the accuracy of intrusion event classification.

5. Conclusions

In response to issues such as unclear characteristics of intrusion events in drainage tunnels, unknown effects of feature parameters on intrusion event classification, and the potential restriction of high-dimensional feature parameters on the accuracy of intrusion event classification, research was conducted based on PCA and neural networks. The main conclusions are as follows:

(1) A monitoring system based on DAS was established for intrusion event simulation in drainage tunnels. Typical characteristics of events such as rockfalls and water release were obtained. However, due to the complexity of the intrusion signals in the drainage tunnels, it is difficult to accurately classify intrusion events using only one or a few feature parameters.
(2) Principal component analysis indicated relatively significant correlations among feature parameters, necessitating the use of PCA to mitigate these correlations. Features related to signal amplitude, such as maximum amplitude, mean, and energy, contributed substantially to the DAS signal classification, followed by signal entropy. The first four principal components account for 48.71% of the variance, and there is a 54.8% reduction in the dimensionality of the feature parameters.
(3) The classification accuracy of intrusion events using PCA-processed data is higher than that without PCA processing, with overall recognition accuracies for rockfall events and water releases at 79.1% and 72.7%, respectively. Moreover, the performance of the artificial neural network model surpassed that of Bayesian and logistic regression models, demonstrating ANNs' advantage in handling complex models such as intrusion event classification.

Author Contributions: Methodology, X.S.; software, X.S.; writing—original draft preparation, X.S.; funding—P.Y.; supervising—P.Y.; resources, P.Y.; writing—review and editing, W.Z. and Y.P. All authors have read and agreed to the published version of the manuscript.

Funding: This research was funded by the China Yangtze Power Co., Ltd. Technology Development Project under Grant Z422302020.

Data Availability Statement: The data presented in this study are available on request from the corresponding author. The data are not publicly available due to the project confidentiality.

Conflicts of Interest: Authors Peng Yuan and Weihao Zhang are employed by the company China Yangtze Power Co., Ltd. The remaining authors declare that the research was conducted in the absence of any commercial or financial relationships that could be construed as a potential conflict of interest.

References

1. Yi, C.; Senent, S.; Jimenez, R. Effect of Advance Drainage on Tunnel Face Stability Using Limit Analysis and Numerical Simulations. *Tunn. Undergr. Space Technol.* **2019**, *93*, 103105. [CrossRef]
2. Zou, Q.; Chen, Z.; Zhan, J.; Chen, C.; Gao, S.; Kong, F.; Xia, X. Morphological Evolution and Flow Conduction Characteristics of Fracture Channels in Fractured Sandstone under Cyclic Loading and Unloading. *Int. J. Min. Sci. Technol.* **2023**, *33*, 1527–1540. [CrossRef]

3. Gu, H.; Lai, X.; Tao, M.; Momeni, A.; Zhang, Q. Dynamic Mechanical Mechanism and Optimization Approach of Roadway Surrounding Coal Water Infusion for Dynamic Disaster Prevention. *Measurement* **2023**, *223*, 113639. [CrossRef]
4. Pipitone, C.; Maltese, A.; Dardanelli, G.; Lo Brutto, M.; La Loggia, G. Monitoring Water Surface and Level of a Reservoir Using Different Remote Sensing Approaches and Comparison with Dam Displacements Evaluated via GNSS. *Remote Sens.* **2018**, *10*, 71. [CrossRef]
5. Wai-Lok Lai, W.; Dérobert, X.; Annan, P. A Review of Ground Penetrating Radar Application in Civil Engineering: A 30-Year Journey from Locating and Testing to Imaging and Diagnosis. *NDT E Int.* **2018**, *96*, 58–78. [CrossRef]
6. Gorshkov, B.G.; Yüksel, K.; Fotiadi, A.A.; Wuilpart, M.; Korobko, D.A.; Zhirnov, A.A.; Stepanov, K.V.; Turov, A.T.; Konstantinov, Y.A.; Lobach, I.A. Scientific Applications of Distributed Acoustic Sensing: State-of-the-Art Review and Perspective. *Sensors* **2022**, *22*, 1033. [CrossRef]
7. Shang, Y.; Sun, M.; Wang, C.; Yang, J.; Du, Y.; Yi, J.; Zhao, W.; Wang, Y.; Zhao, Y.; Ni, J. Research Progress in Distributed Acoustic Sensing Techniques. *Sensors* **2022**, *22*, 6060. [CrossRef] [PubMed]
8. Xie, T.; Zhang, C.-C.; Shi, B.; Wang, Z.; Zhang, S.-S.; Yin, J. Seismic Monitoring of Rockfalls Using Distributed Acoustic Sensing. *Eng. Geol.* **2023**, *325*, 107285. [CrossRef]
9. Zhang, Y.; Zhao, W.; Dong, L.; Zhang, C.; Peng, G.; Shang, Y.; Liu, G.; Yao, C.; Liu, S.; Wan, N.; et al. Intrusion Event Identification Approach for Distributed Vibration Sensing Using Multimodal Fusion. *IEEE Sens. J.* **2024**. [CrossRef]
10. Stajanca, P.; Chruscicki, S.; Homann, T.; Seifert, S.; Schmidt, D.; Habib, A. Detection of Leak-Induced Pipeline Vibrations Using Fiber—Optic Distributed Acoustic Sensing. *Sensors* **2018**, *18*, 2841. [CrossRef]
11. Fang, N.; Wang, L.; Jia, D.; Shan, C.; Huang, Z. Walking Intrusion Signal Recognition Method for Fiber Fence System. In Proceedings of the 2009 Asia Communications and Photonics conference and Exhibition (ACP), Shanghai, China, 2–6 November 2009; pp. 1–6.
12. Sun, Q.; Feng, H.; Yan, X.; Zeng, Z. Recognition of a Phase-Sensitivity OTDR Sensing System Based on Morphologic Feature Extraction. *Sensors* **2015**, *15*, 15179–15197. [CrossRef] [PubMed]
13. Wang, B.; Pi, S.; Sun, Q.; Jia, B. Improved Wavelet Packet Classification Algorithm for Vibrational Intrusions in Distributed Fiber-Optic Monitoring Systems. *Opt. Eng.* **2015**, *54*, 055104. [CrossRef]
14. Zhao, Y.; Zhong, Z.; Li, Y.; Shao, D.; Wu, Y. Ensemble Empirical Mode Decomposition and Stacking Model for Filtering Borehole Distributed Acoustic Sensing Records. *Geophysics* **2023**, *88*, WA319–WA334. [CrossRef]
15. Xu, C.; Guan, J.; Bao, M.; Lu, J.; Ye, W. Pattern Recognition Based on Enhanced Multifeature Parameters for Vibration Events in φ-OTDR Distributed Optical Fiber Sensing System. *Microw. Opt. Technol. Lett.* **2017**, *59*, 3134–3141. [CrossRef]
16. Wang, X.; Liu, Y.; Liang, S.; Zhang, W.; Lou, S. Event Identification Based on Random Forest Classifier for Φ-OTDR Fiber-Optic Distributed Disturbance Sensor. *Infrared Phys. Technol.* **2019**, *97*, 319–325. [CrossRef]
17. Wu, H.; Qian, Y.; Zhang, W.; Tang, C. Feature Extraction and Identification in Distributed Optical-Fiber Vibration Sensing System for Oil Pipeline Safety Monitoring. *Photonic Sens.* **2017**, *7*, 305–310. [CrossRef]
18. Xu, C.; Guan, J.; Bao, M.; Lu, J.; Ye, W. Pattern Recognition Based on Time-Frequency Analysis and Convolutional Neural Networks for Vibrational Events in φ-OTDR. *Opt. Eng.* **2018**, *57*, 016103. [CrossRef]
19. Li, Z.; Zhang, J.; Wang, M.; Zhong, Y.; Peng, F. Fiber Distributed Acoustic Sensing Using Convolutional Long Short-Term Memory Network: A Field Test on High-Speed Railway Intrusion Detection. *Opt. Express* **2020**, *28*, 2925–2938. [CrossRef]
20. Grosse, C.U. *Acoustic Emission Testing: Basics for Research-Applications in Civil Engineering*, 1st ed.; Grosse, C., Ohtsu, M., Eds.; Springer: Berlin/Heidelberg, Germany, 2008; ISBN 978-3-540-69895-1.
21. Wanliss, J.A.; Wanliss, G.E. Efficient Calculation of Fractal Properties via the Higuchi Method. *Nonlinear Dyn.* **2022**, *109*, 2893–2904. [CrossRef]
22. Lever, J.; Krzywinski, M.; Altman, N. Principal Component Analysis. *Nat. Methods* **2017**, *14*, 641–642. [CrossRef]
23. Reyes, J.; Morales-Esteban, A.; Martínez-Álvarez, F. Neural Networks to Predict Earthquakes in Chile. *Appl. Soft Comput.* **2013**, *13*, 1314–1328. [CrossRef]
24. Tharwat, A. Principal Component Analysis—A Tutorial. *Int. J. Appl. Pattern Recognit.* **2016**, *3*, 197–240. [CrossRef]
25. Abdolrasol, M.G.M.; Hussain, S.M.S.; Ustun, T.S.; Sarker, M.R.; Hannan, M.A.; Mohamed, R.; Ali, J.A.; Mekhilef, S.; Milad, A. Artificial Neural Networks Based Optimization Techniques: A Review. *Electronics* **2021**, *10*, 2689. [CrossRef]
26. Geiger, D.; Verma, T.; Pearl, J. Identifying Independence in Bayesian Networks. *Networks* **1990**, *20*, 507–534. [CrossRef]

water

MDPI

Article

Monitoring and Evaluation of Debris Flow Disaster in the Loess Plateau Area of China: A Case Study

Baofeng Wan *, Ning An and Gexue Bai

Gansu Institute of Engineering Geology, Lanzhou 730000, China; ann8299089@163.com (N.A.); baigexue6009@163.com (G.B.)
* Correspondence: w198611052023@163.com

Abstract: The Loess Plateau area, with complex geomorphological features and geological structure, is highly prone to geologic disasters such as landslides and debris flow, which cause great losses. To investigate the initiation mechanism of landslide and debris flow disasters and their spreading patterns, historical satellite images in the Laolang gully were collected and digitized to generate three-dimensional topographic and geomorphological maps. Typical landslides were selected for landslide thickness measurement using a standard penetrometer and high-density electrical method. Numerical models were established to simulate the occurrence and development of landslides under different working conditions and to evaluate the spreading range based on the propagation algorithm and friction law. The results show that the 10 m resolution DEM data are well matched with the potential hazard events observed in the field site. The smaller the critical slope threshold, the greater the extent and distance of landslide spreading. The larger the angle of arrival, the greater the energy loss, and therefore the smaller the landslide movement distance. The results can provide scientific theoretical guidance for the prevention and control of rainfall-induced landslide and debris flow disasters in the Loess Plateau area.

Keywords: debris flow disaster; risk assessment; Loess Plateau area

Citation: Wan, B.; An, N.; Bai, G. Monitoring and Evaluation of Debris Flow Disaster in the Loess Plateau Area of China: A Case Study. *Water* **2024**, *16*, 2539. https://doi.org/10.3390/w16172539

Academic Editors: Helong Gu, Xueyi Shang and Huatao Zhao

Received: 11 June 2024
Revised: 24 August 2024
Accepted: 29 August 2024
Published: 8 September 2024

1. Introduction

As one of the most catastrophic geologic disasters [1,2], debris flow disasters cause serious economic losses and casualties every year, posing a great threat to socio-economic development and the safety of human life and property [3,4]. Debris flow monitoring is a prior means of the related research, which provides the basis for its theoretical study, experimental study, mechanism analysis, physical process analysis, and mathematical simulation, as well as the basis for debris flow early warning. Numerous scholars and research institutions that have conducted systematic research on the basic conditions and excitation factors of debris flow formation [5–8] have put forward the basis of discrimination of mudslide occurrence, carried out mudslide monitoring, and established the discrimination model of mudslides.

The occurrence of debris flow is closely related to the stability of the slope. The failure of a slope easily induces debris flow disasters. Therefore, slope stability analysis and debris flow disaster assessment are important research branches. Wahab et al. [1] conducted an experiment on a physical model by demonstrating the debris flow at different slope angles and provided a risk assessment approach to mitigate the impact regions of debris flow. Zhou et al. [3] selected 17 influencing factors to establish a hazard assessment model using a machine learning algorithm, which could provide support for the prevention of debris flow hazards. Rajabian and Shukla [9] developed an analytical approach based on the friction circle method to predict the safety factor of anchor-reinforced slopes using Taylor's chart. The predicted results were in good agreement with the results of the limit equilibrium method. Dong et al. [10,11] established two innovative interval non-probabilistic reliability evaluation methods for the surrounding jointed rockmass, in which the uncertainty of the rockmass parameters were eliminated using the interval theory. Then, the established methods were successfully applied to the slope stability analysis of a tailings dam and achieved good results [12]. Charles et al. [13] proposed a new three-dimensional (3D) theoretical model to describe the effect of hydraulic mechanical reinforcement of the root system on the stability of initially unsaturated soil slopes. Ni et al. [14] investigated the effect of plant reinforcement on the stability of coarse-grained soil slopes and pointed out that the mechanical enhancement was only effective in the upper layers. Qu et al. [15] investigated the reliability of unsaturated soil slopes under rainfall infiltration using a 3D saturated water permeability rotating anisotropic random field. Subramanian et al. [16] recommended a two-dimensional numerical modeling approach for soil slope stability assessment considering the water content changes of the soil induced by seasonal climatic effects. Freeze–thaw action and snowmelt water infiltration were found to have considerable impacts on soil slope stability. Alberto et al. [17] determined the spatial relationship between landslides and glaciers and established a preliminary relative chronology. Pierson [18] summarized the kinematic, volumetric, and hydraulic characteristics of ten large historic volcanic debris flows from four different volcanoes. Fang et al. [19] introduced a self-similar percolation model and developed it from two dimensions to three dimensions, which explained why the joint on the other side of the block with the main river is easier to break than other places of the block. Gao and Zhang [20] comprehensively considered both the frequency and scale of landslides to analyze the landslide susceptibility of a small watershed. Chen et al. [21] performed a model study to debris flow and proposed the mechanisms of initiation of debris flow.

Scientific early warning of debris flow disasters is based on predictions before debris flow initiation [22–24] and warning in the movement [25–27]. The early warning model research is roughly divided into two categories; one establishes mathematical models through data analysis to formulate the critical rainfall of mudslides, and the other establishes early warning models through the analysis of the debris flow formation mechanism. Dai et al. [28] discussed the applicability of a variety of approaches to assessing landslide risk and proposed a landslide risk assessment and management framework for reducing landslide risk. For the early warning studies of critical rainfall, the research mainly focuses on determining the critical rainfall for mudslides and establishing a mudslide early warning model after statistically analyzing the rainfall characteristics of the region where the mudslides are located. After analyzing the data of mudslide occurrence in different places, many scholars have established a series of mudslide early warning models under different conditions of rainfall characteristics [29–34]. For the early warning research on the debris flow formation mechanism, it is mainly based on the critical conditions of debris flow initiation, to study the early warning indicators under different initiation critical conditions, and then establish debris flow early warning models. Timilsina et al. [35] analyzed the soil moisture patterns and evaluated the importance of soil moisture in predicting debris flow initiation sites. The results indicated that the effect of the soil moisture pattern on the initiation of debris flow is smaller than that of the vegetation and slope. Chatra et al. [36] conducted finite difference analysis of transient water flow through unsaturated–saturated soil to analyze the effects of rainfall intensity and duration on slope stability during rainfall events. The results showed that rain-induced instability is much greater in a loose soil slope than in a medium soil slope. Lourenço et al. [37] examined the relationship between soil particle wettability and slope processes in physical models and found that a continuous capping effect generated by water repellency was a necessary condition for the generation of runoff-initiated debris flow. Based on real-time monitoring and numerical simulation, Dong et al. [38] constructed a real-time pre-alarm system using the Internet of Things and cloud computing, through which the stable or dangerous warning signals of human-caused debris flow in a tailings dam can be obtained. Hu et al. [39] performed a series of flume tests to study how the initial soil moisture influences the initiation of debris flow and pointed out that the increase in pore pressure leading to soil liquefaction and the loss of the cohesive strength of the soil are two main triggering mechanisms of debris flows. He et al. [40] analyzed the disaster mechanism and evolution process of the landslide-debris-flow geohazard chain in areas with strong earthquakes through the consideration of the accumulative sliding amount and the critical rainfall. In their study, the influence of strong earthquake activities and rainfall caused the formation of the Niumian Valley geohazard chain. Niu et al. [41] applied the disaster chain theory to study preventive measures for mudslide disasters originating from source wells. They found that the type of disaster chain is a complex cyclical loop and proposed several proactive and reactive control methods.

Loess usually has high porosity, and its pore structure is more sensitive to water. Loess shows obvious plastic characteristics after absorbing water, and it becomes more susceptible to deformation and flow. In the process of drying and wetting, loess will undergo large volume changes, showing obvious expansion and contraction characteristics. As a result, it is susceptible to destabilizing landslide disasters. The Loess Plateau, as the main occurrence area of debris flow [20], has a unique combination of loess gully landforms and inter-valley landforms, such as loess, beams, and mounts. The very fragmented surface and serious soil erosion provide a rich material source base and conditions for the occurrence of debris flow, making the Loess Plateau the most serious area subject to mudslide disasters in China [42]. Therefore, systematic monitoring and early warning research on debris flow in the Loess Plateau area is particularly important and the most cost-effective means to reduce the damage caused by debris flow disasters.

2. Engineering Background and Research Methods

Gansu Province is located at the crossroads of the Qinghai–Tibet Plateau, the Inner Mongolia Plateau, and the Loess Plateau, with complex geological and environmental conditions, fragile ecology, concentrated rainfall, and frequent torrential rains. As a result of these influences, there are frequent geologic hazards in the region, and it is one of the provinces with the most serious geologic hazards in the country. In recent years, mudslide disasters have further aggravated the trend; in just five years after 2010, four consecutive large-scale mudslide disasters occurred, with a large hazardous range and high numbers of victims and economic losses. With the increase in extreme weather, the disaster situation is very serious, and the task of disaster prevention and mitigation is onerous.

Lanzhou is located in the northeastern edge of the Qinghai–Tibet activity block with strong neotectonic activity, in the composite part of the north–south seismic zone and Qilian Mountain seismic zone, with complex geological structure, strong neotectonic movement, and special topography and geomorphology, resulting in a narrow urban space. Coupled with the increasing intensity of human engineering activities and the frequent occurrence of extreme weather and climate events, the risk of geologic hazards is very serious. Lanzhou is deep inland, located in the transition zone between the monsoon climate zone and non-monsoon climate zone. According to the statistics of meteorological stations in Lanzhou City and neighboring districts, the average precipitation in Lanzhou City has been 300–600 mm for many years, the distribution of precipitation in the territory is very uneven, the regional precipitation decreases from the south to the north, the precipitation along the Qidaoliang area of the southern mountainous region is over 550 mm, and the precipitation along the Sangyuanzi area of the northeastern part of the country is over 550 mm. In the northeast, the precipitation is 250 mm in the area of Sangyuanzi, and under the influence of the monsoon, the distribution of precipitation in each month of the year is also very uneven in Lanzhou. Geological disasters in Lanzhou City are dominated by landslides, avalanches, and mudslides, with landslides and avalanches developed in a band-like and piecemeal manner, concentrated at the front of the mountains on both sides of the Yellow River and the front edge of the high terraces of the river valleys, while mudslides are distributed in a line-like manner along the riverbanks. As of the end of September 2015, the city had identified 2446 geologic hazard sites, including 346 landslides, 194 avalanches, 1639 unstable slopes (potential avalanches and landslides), 255 mudslides, and 12 ground collapses.

The Laolang gully is located in the south side of Lanzhou City, east of Gaolan Mountain, north of Minzu Village, Gaolan Mountain Township, Chengguan District, with geographic coordinates E 103°50′30″–103°51′6″, N 36°0′19″–36°1′31″. The study area is 950 m away from Lanzhou Station, and there are a large number of public facilities and residential areas near the area, and if a debris flow disaster occurs, it will seriously threaten the local transportation, economy, and the safety of residents' lives and properties. There have been 91 loess landslides in the Laolang gully, and the loose piles generated by landslides and collapses provide sufficient material sources for the occurrence of mudslides. To scientifically prevent the occurrence of debris flow disasters in the Laolang gully, according to the spatial and temporal distribution characteristics of loess landslides and debris flows in the Laolang gully, meteorological observation, landslide observation, debris flow observation, and image observation systems are arranged in the debris flow monitoring area.

In this study, satellite images were used to circle the landslide extent and combined with DEM data to realize the three-dimensional visualization of the study area to further improve the accuracy of mudslide source investigation and analysis. The historical satellite images of the study area were first collected, and in order to explore the influence of external factors on the quality of satellite images, they were organized according to different shooting climates, month, time, and other factors, and compared and analyzed. Then, in order to realize the three-dimensional visualization of the study area, the 1:10,000 topographic maps of the study area were digitized in ArcGIS, and the DEM data were generated and combined with selected satellite images to generate three-dimensional topographic geomorphological maps, which were then interpreted, and the circled landslide extent was verified through on-site investigations. The landslides in the study area were categorized through field surveys, and there are two types of movement: sliding and collapse. The sliding movement includes rotational landslides and translational landslides. A rotational landslide refers to the movement of a sliding body along a sliding surface that curves upward, while a transitional landslide involves downward and lateral movement along a relatively flat sliding surface. Rotational landslides often occur on steep slopes with relatively uniform soil, where the sliding surface is curved, causing the landslide to rotate. Transitional landslides typically occur in rock layers, especially in the presence of weak layers, where the sliding surface is flat and the landslide mass moves along a plane. In the engineering field, the distinction can be made based on the following methods. Firstly, the shape of the two landslides are different: rotational landslides are usually curved and shaped like a semicircle, whereas translational landslides are flatter and tend to be straight or slightly inclined. Secondly, the modes of movement of the two landslides are different: rotational landslides rotate and slide around a point, whereas translational landslides slide along a plane. Finally, the surface characteristics of the two landslides are different: rotational landslides may have depressions and bulges in their surface, whereas translational landslides generally have a more regular surface. There were 91 landslides in the monitoring area, of which the numbers of rotational landslides, translational landslides, and collapses were 49, 37, and 5, respectively. The distribution of the landslides is shown in Figure 1a. Rotational landslides and translational landslides accounted for 54% and 41% of the total number of landslides, respectively, and became the main types of landslides in the study area. Therefore, this study focused on the distribution characteristics and morphological features of rotational landslides and translational landslides.

To measure the landslide thickness, 19 representative landslides in the study area were selected for thickness measurement using a standard penetrometer, including 9 rotational and 10 translational landslides, which were uniformly distributed in different locations in the study area, as marked in Figure 1b. Use of a standard penetrometer is a common geotechnical engineering method to detect soil strength and divide strata. The hammer is controlled to hit the interceptor nut installed on the drill rod by manpower or automatic device, so that the drill rod penetrates into the soil layer at a certain depth, and then the soil strength parameter of the corresponding depth is obtained with the measured penetration hammer number N, and the soil strength parameters of the corresponding depth are calculated. To verify the reliability of the standard penetrometer method for determining the thickness of landslides, two landslides, as shown in Figure 1b, were selected for high-density electrical method measurement in this study. The measurements were compared with those of the standard penetrometer, and the reliability of the standard penetrometer method was examined on the basis of the degree of difference between the two methods. The high-density resistivity method combines the characteristics of electrical profiling and electrical sounding. Typically, it involves an A electrode and a B electrode supplying power, while an M electrode and an N electrode receive the potential difference to measure and calculate apparent resistivity and polarization. During on-site measurements, all electrodes are arranged at intervals on designated measurement points, with the main unit automatically controlling the variations in supply and receiving electrodes to complete the measurements. This paper selected the "dipole–dipole" measurement method to determine the thickness of the sliding body and utilized RES2DINV 5.0 software for data inversion, further acquiring detailed information on the sliding surface.

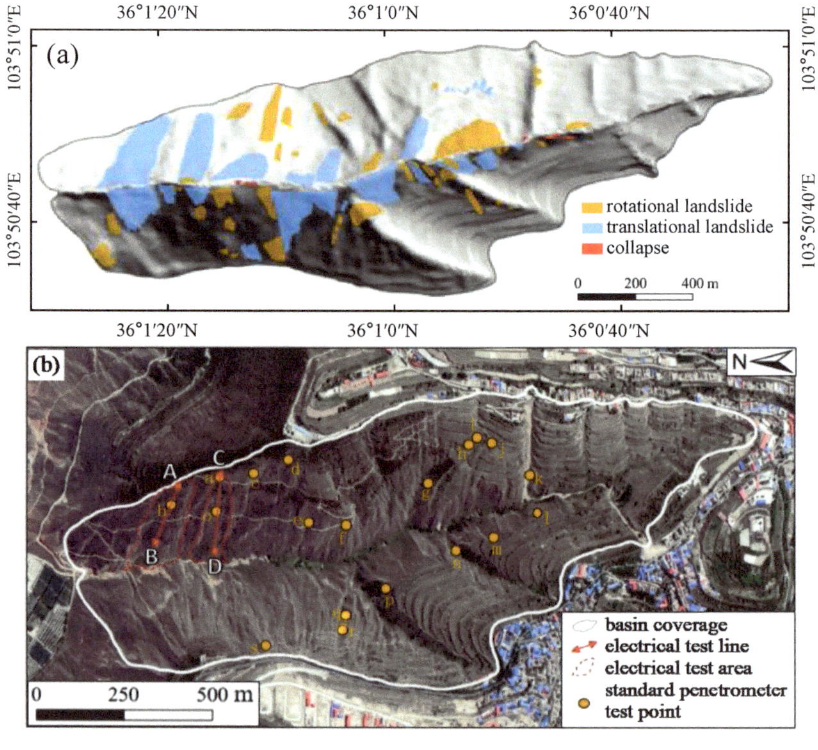

Figure 1. (**a**) The distribution of the landslides; (**b**) Thickness measurements using a standard penetrometer and high-density electrical method; The uppercase letters A to D are the boundaries of the electrical test line; The lowercase letters a to s are the standard penetrometer test point, respectively.

The standard penetration method can obtain the landslide thickness at the monitoring point with high accuracy, but it is only the result of a single point measurement, and the landslide thickness in the neighboring area needs to be inferred from the measurement result at the monitoring point. Conversely, the high-density electrical method can obtain the landslide thickness in the area, but the accuracy is lower than that of the standard penetration method. Numerical simulation methods can simulate the dynamic behavior of mudslides, predicting their propagation paths and influence range, which is of great significance for disaster risk assessment and the design of early warning systems. In this

of the main gully extending in the N–S direction. The debris formed by avalanches affects the topography of the local area to a certain extent, and at the same time, these debris can be regarded as a good source of debris flow hazard. It is believed that the debris on both sides of the channel provides part of the energy for the movement in the mudslide channel, and with the frictional consumption of the movement process, the energy is gradually dissipated at the mouth of the channel, and the movement stops.

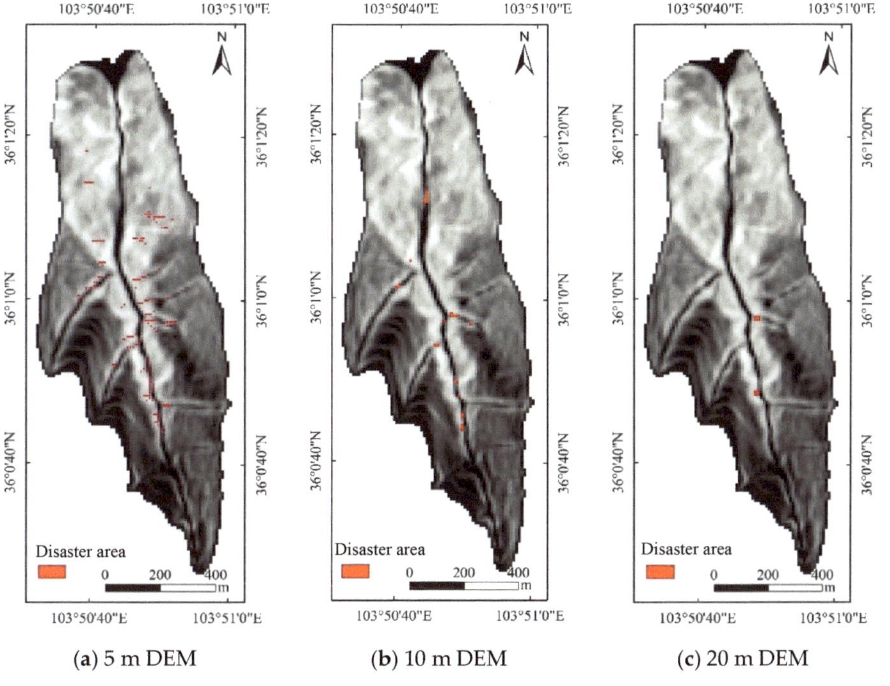

Figure 4. Disaster area distributions under flood conditions identified in DEM data at different resolutions.

Analyzing the spreading results of landslide-influenced debris flow with different precision DEM data, as shown in Figures 8 and 9, the comparison revealed that the simulation results at 10 × 10 m resolution were more reasonable compared with the results at other resolutions. After the field engineering geomorphological mapping and remote sensing image interpretation, it was found that large numbers of landslides were distributed on the slopes on both sides of the gully in the study area, the number of landslides at the mouth of the gully was significantly higher than that at the end of the gully, the landslides were mainly distributed in the gully entrance area where the gradient is steeper, but most of them were mainly small-sized landslides, and the tail area was slower due to the slope gradient and the vegetation cover was higher, so there were fewer landslides on the steep slopes with less water than those at the mouth of the gully, but at larger scales. The occurrence of landslides is often accompanied by the formation of large-scale joint cracks, which has a greater impact on the local hydrological conditions. At the same time, the topography and geomorphology have certain roles in changing the local gradient and catchment conditions of the slope. Therefore, affected by a large number of landslides on the slopes on both sides of the channel, the debris flow events in the watershed were re-identified and simulated. Landslides on the slopes provide large numbers of material sources for debris flow, and the slopes on both sides of the central part of the channel are flat with no obvious tributary ditch development, so that the expansion of the material source movement on both sides has a larger range.

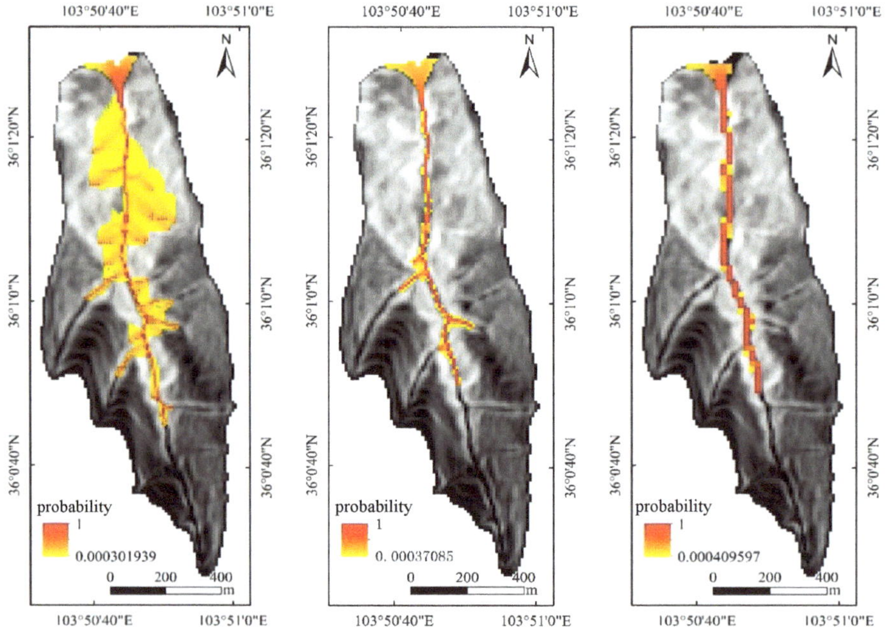

(**a**) 5 m DEM (**b**) 10 m DEM (**c**) 20 m DEM

Figure 5. Spreading results under flood conditions identified in DEM data at different resolutions.

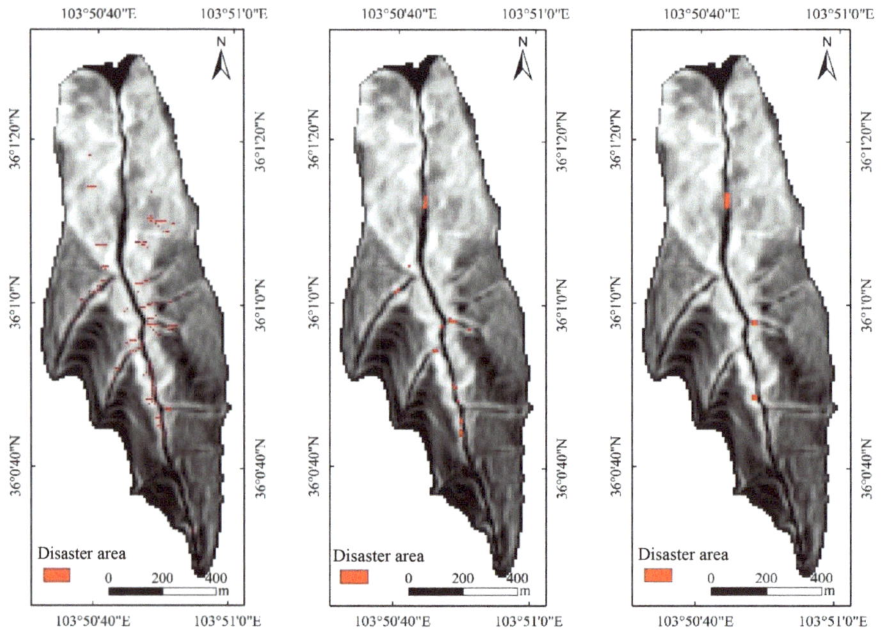

(**a**) 5 m DEM (**b**) 10 m DEM (**c**) 20 m DEM

Figure 6. Disaster area distributions under collapse-affected debris flow conditions identified in DEM data at different resolutions.

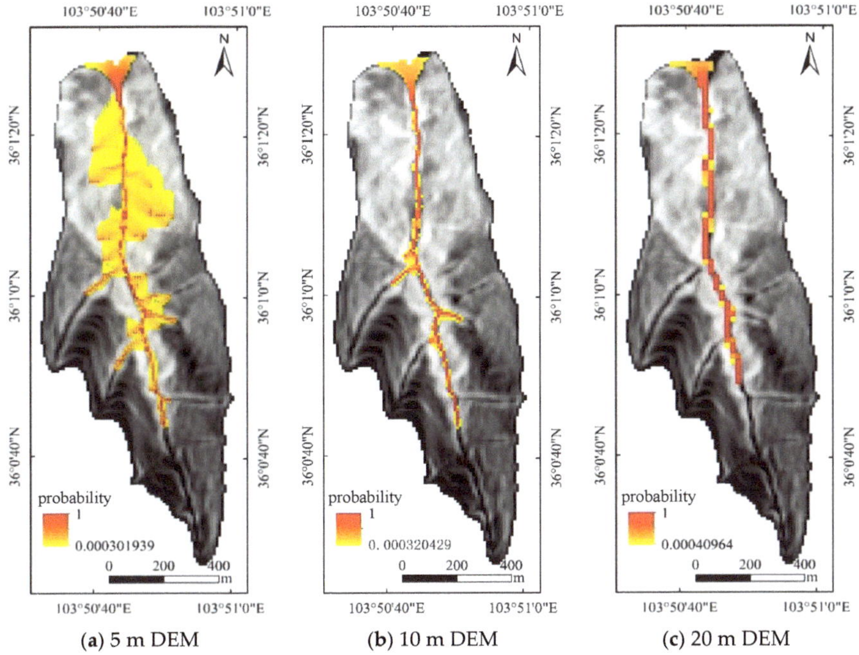

Figure 7. Spreading results under collapse-induced debris flow conditions identified in DEM data at different resolutions.

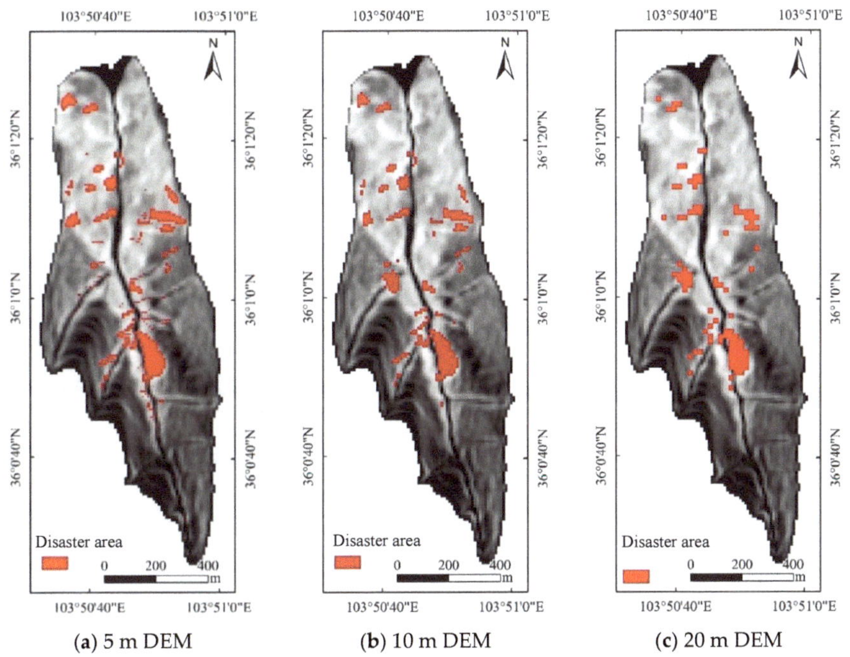

Figure 8. Disaster area distributions under landslide-induced debris flow conditions identified in DEM data at different resolutions.

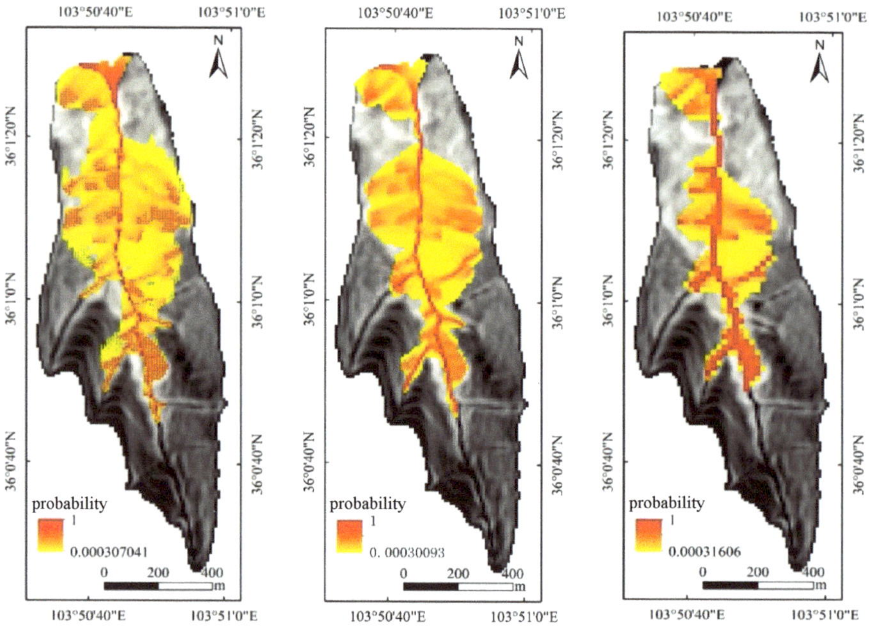

(**a**) 5 m DEM (**b**) 10 m DEM (**c**) 20 m DEM

Figure 9. Spreading results under landslide-induced debris flow conditions identified in DEM data at different resolutions.

Under extreme climatic conditions, landslides and collapse usually occur at the same time. In order to analyze the occurrence of debris flow under extreme conditions, both collapse and landslide were considered as potential sources. Based on the landslide source area, the collapse disaster area at the head of the upstream gully was further identified. It was analyzed that the generation of large numbers of material sources provides great kinetic energy for the movement of debris flow. Compared with the debris flow movement process only considered to be affected by landslides, the debris flow movement process under such extreme conditions still has large energy at the mouth of the trench, and the simulation results are shown in Figures 10 and 11.

By simulating the identification of source areas under four different scenarios of flood, collapse-induced debris flow, landslide-induced debris flow, and debris flow under the simultaneous influence of collapse and landslide, it was found that the DEM data with a resolution of 5 m were the most refined, which not only identified the potential source areas on both sides of the near-channel but also mistakenly identified many small-scale topographic undulations on the slope surface as source areas, resulting in simulated movement results that did not correspond to the actual situation. The DEM data with a precision of 20 m were too coarse, and the potential source areas in some branch channels were not well identified, and many small-scale source areas were ignored. The disaster area identified by the 10 m resolution DEM data, although still somewhat different from the actual situation, was the most consistent result compared to the 5 m and 20 m resolution results.

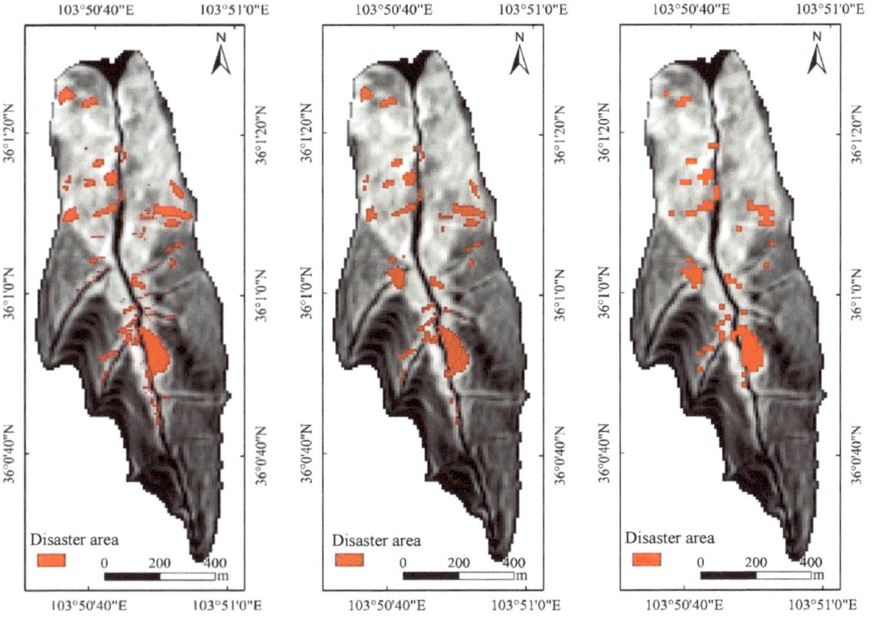

(**a**) 5 m DEM (**b**) 10 m DEM (**c**) 20 m DEM

Figure 10. Disaster area distributions under the simultaneous influence of collapse and landslide conditions identified in DEM data at different resolutions.

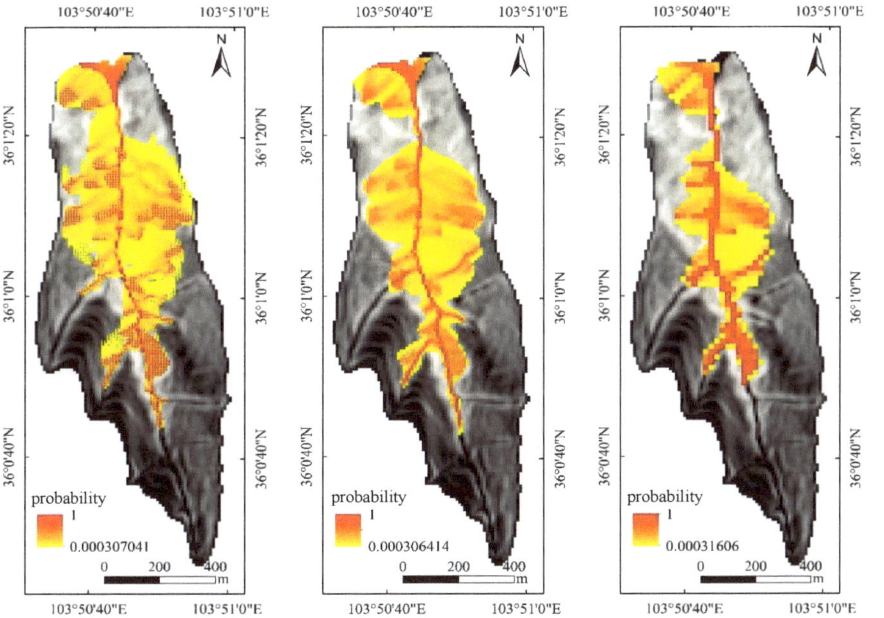

(**a**) 5 m DEM (**b**) 10 m DEM (**c**) 20 m DEM

Figure 11. Spreading results under the simultaneous influence of collapse and landslide conditions identified in DEM data at different resolutions.

3.3. Risk Assessment of Debris Flow at Different Critical Slope Thresholds

The effect of critical terrain slope for source area identification was analyzed with the 10 m resolution DEM data. Three critical terrain slope thresholds of 15°, 25°, and 35° were analyzed separately to identify flood source areas without considering the effect of landslides from avalanches. It was found that areas smaller than this slope were excluded from the potential source area due to the increase in the source area initiation slope threshold, as shown in Figures 12 and 13. Thus, as the critical terrain slope for source area identification increased, the identified source areas gradually decreased. In addition to some of the source areas in the upstream branches of the main ditch, some of the source areas in the middle and lower reaches of the ditch were also controlled by topographic features and could not be well identified when the source area initiation slope threshold was large. When the source area initiation slope threshold increased, the material that can undergo flow decreased, and the range and distance of movement spreading were relatively reduced. The evaluated disaster-prone area may be too small, and the obtained hazard evaluation was not reasonable.

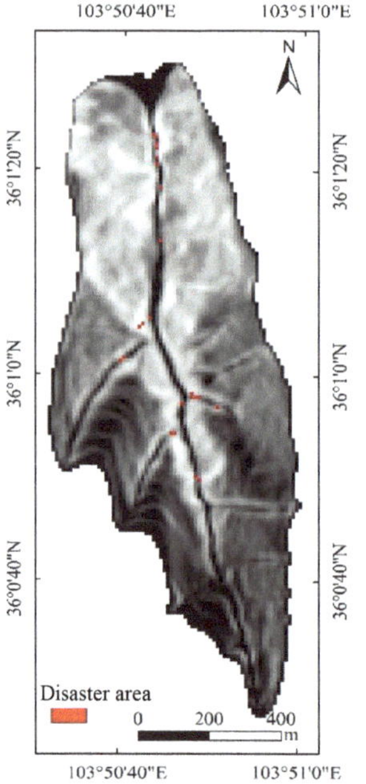

(a) slope angle 15°

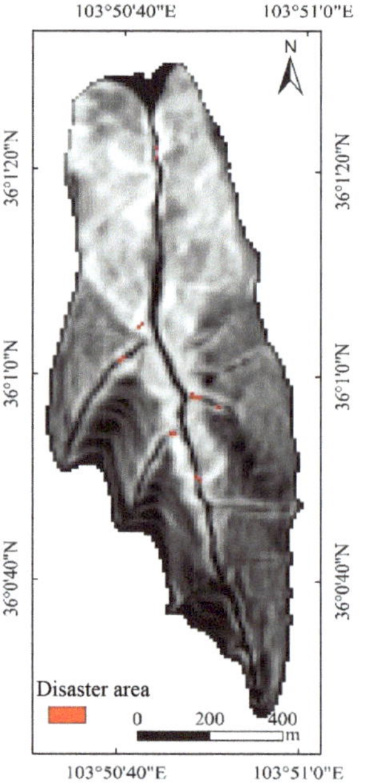

(b) slope angle 25°

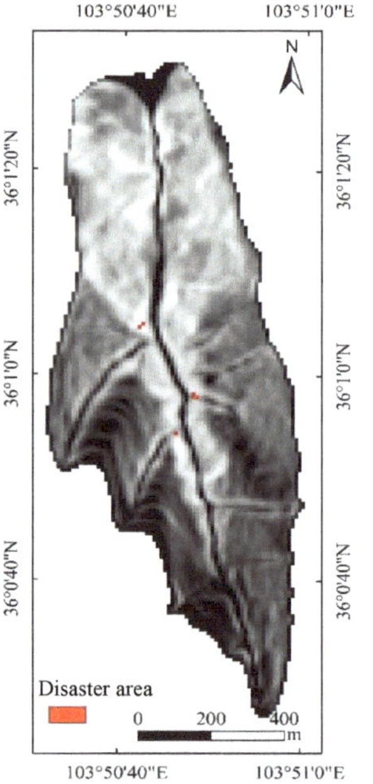

(c) slope angle 35°

Figure 12. Disaster area distributions under flood conditions at different critical slope thresholds.

Three critical terrain slope thresholds of 15°, 25°, and 35° were analyzed to identify debris flow source areas considering the combined effects of collapse and landslide. As the topographic slope threshold increased, the range of material sources at the identification site on the slope decreased, and some source areas larger than the topographic slope threshold could not be identified, as shown in Figures 14 and 15. Therefore, the closest potential source accumulations need to be identified by selecting the appropriate terrain slope threshold based on the potential sources investigated in the field. The reduction in physical sources causes the debris flow material in the trench to have less kinetic energy of movement, the energy to reach the trench entrance and exit quickly disappears, and the distance of movement becomes shorter.

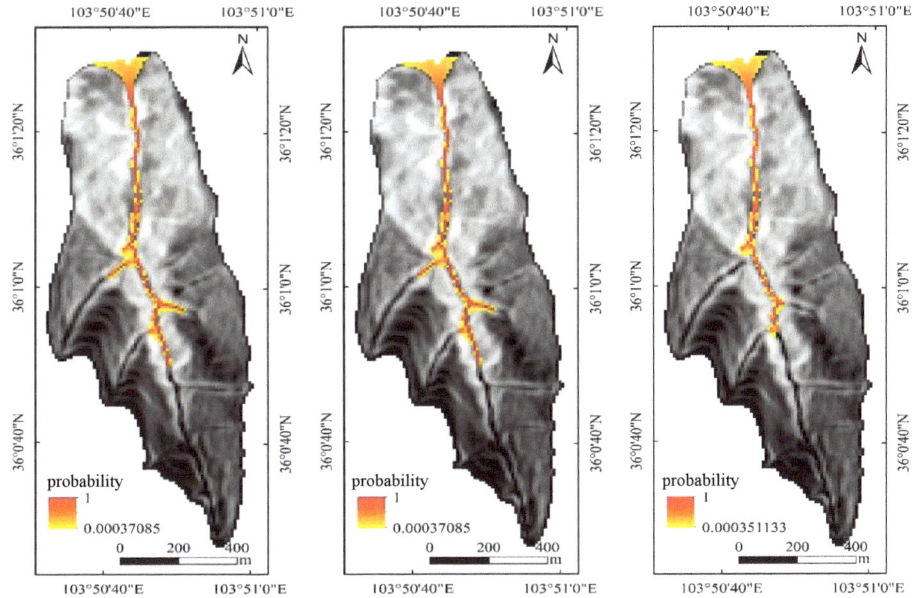

(a) slope angle 15° (b) slope angle 25° (c) slope angle 35°

Figure 13. Spreading results under flood conditions at different critical slope thresholds.

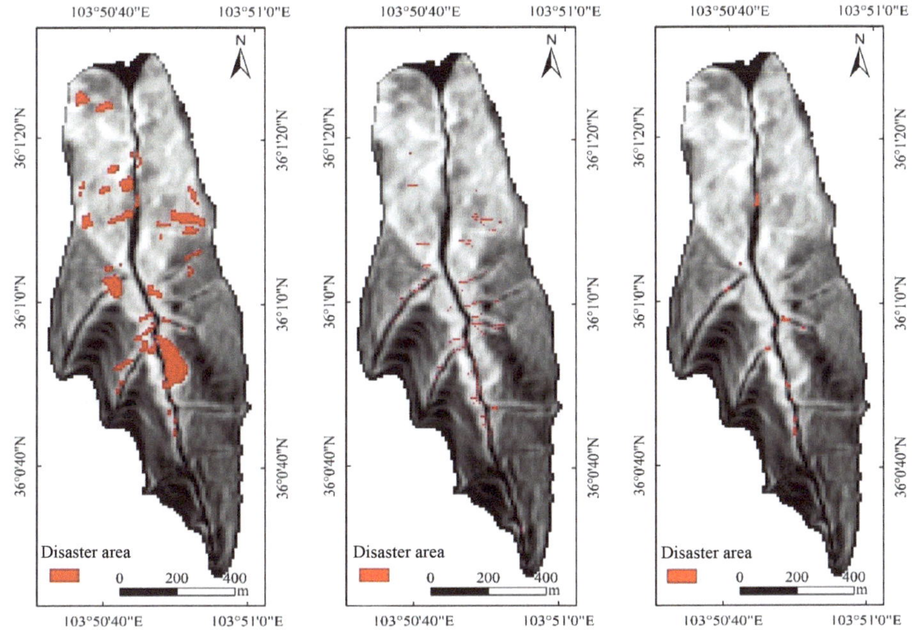

(a) slope angle 15° (b) slope angle 25° (c) slope angle 35°

Figure 14. Disaster area distributions under the simultaneous influence of collapse and landslide conditions at different critical slope thresholds.

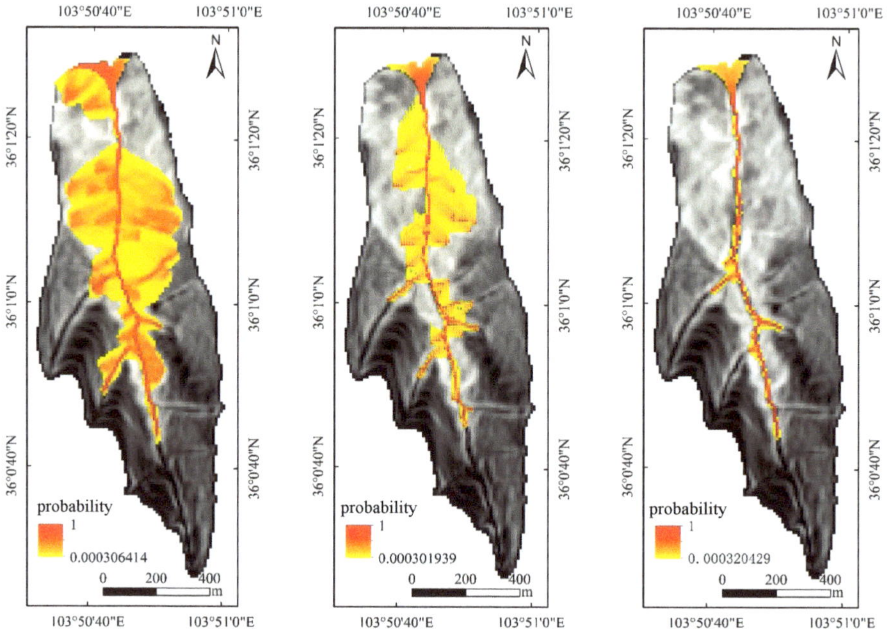

(a) slope angle 15° (b) slope angle 15° (c) slope angle 15°

Figure 15. Spreading results under the simultaneous influence of collapse and landslide conditions at different critical slope thresholds.

3.4. Risk Assessment of Debris Flow under the Spreading of Different Substances

The motion paths and distances of the same material source are often constrained by potential energy, topography, and other conditions, even if the same material source area often has different motion diffusion paths and diffusion ranges under different conditions. The motion diffusion path in the model is mainly controlled by the index x in the Holmgren model, and the four values of x were selected as 2, 4, 6, and 8, respectively, to analyze the influence of the index x in the Holmgren model on the motion process for the raster data with 10 m resolution. Figure 16 shows the spreading process of debris flow under the influence of both collapse and landslide, and it was found that the spreading range tended to decrease and the flow rate converge rapidly with the increase in the index x. Figure 17 shows the spreading range of motion of the flood, and it was similarly found that as the exponent x became smaller, its spreading range to both sides of the channel became wider. The two results reflected that the index x in the Holmgren model mainly affected the degree of diffusion to both sides during the movement process, and when the index x was larger, the diffusion range was larger, all other conditions being equal. The index x in the model is an indirect reflection of the representation of the way of motion diffusion. The model is based on a simple energy balance and, to a certain extent, can represent the energy of the moving material, and so the greater the energy of the moving material, the wider the path of motion, with a greater range of diffusion.

For the flood disaster area identified at 10 m resolution, the difference in its motion process simulated by changing the angle of arrival parameter in the simplified friction-limited model (SFLM model) was determined to analyze the effect of the motion distance of this parameter. The simplified friction limit model is based on the maximum possible distance of motion and is characterized by the smallest travel angle, also known as the angle of arrival, which is the angle between the connection line and the horizontal line between the source area and the farthest point reached by the debris flow along its path.

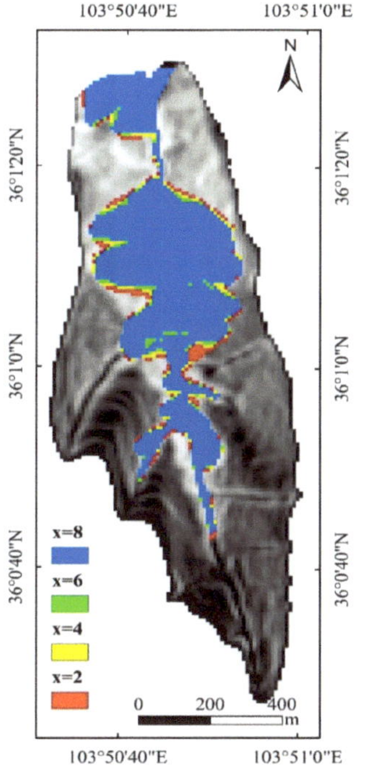

Figure 16. Spreading results under the simultaneous influence of collapse and landslide conditions at different index x values.

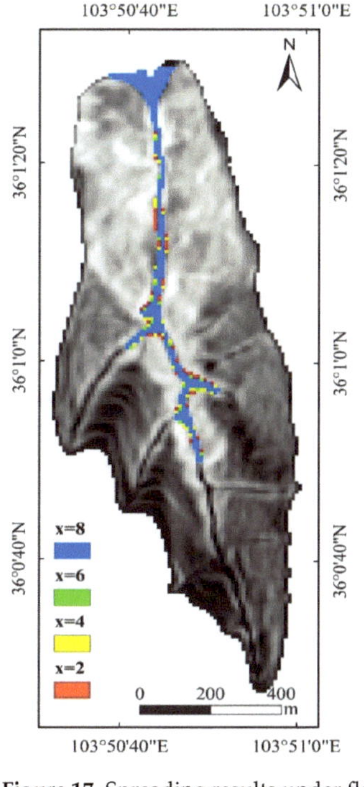

Figure 17. Spreading results under flood conditions at different index x values.

Because a large number of international studies have concluded that the value of this parameter is mostly in the range of 7~11°, three angles of 7°, 9°, and 11° were selected for comparative analysis, and the simulated flood hazard evaluation under the consideration of rainfall surface runoff only is shown in Figure 18. It was found that the maximum movement distance decreased as the angle increased. It was found that when the angle of arrival was set to 11°, the maximum movement distance had not reached the gully entrance and the movement process had stopped. By contrast, when the angle of arrival was set to 7°, the kinetic energy at the mouth of the ditch was not yet fully dissipated and would move further northward out of the ditch until the kinetic energy was reduced to 0, stopping the flow buildup to form a stacked fan. The results of the debris flow hazard evaluation under the consideration of the impact of landslide of avalanche are shown in Figure 19. Large numbers of material sources provided sufficient energy, and the debris flow still flowed out of the gully even when the arrival angle was set to 11°, but the comparison revealed that the energy at the mouth of the gully decreased compared to that at 7° and 9°, and it could be deduced that after flowing out of the gully, it had the shortest distance to move. Since the SFLM model is based on a simplified energy balance, the larger angle represents that the energy loss due to friction is also serious. Therefore, this parameter has a certain physical basis and has better practicality. When the particles of the debris material are rough and the friction between the particles is enhanced, the angle tends to be larger and the distance of movement is shorter.

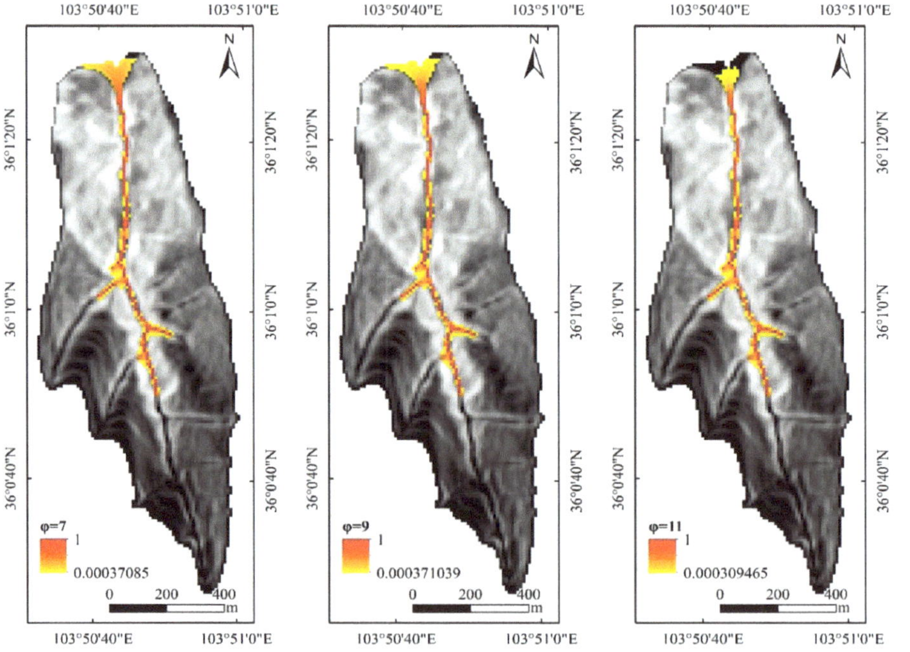

(a) φ = 7° (b) φ = 9° (c) φ = 11°

Figure 18. Spreading results of flooding under different arrival angles.

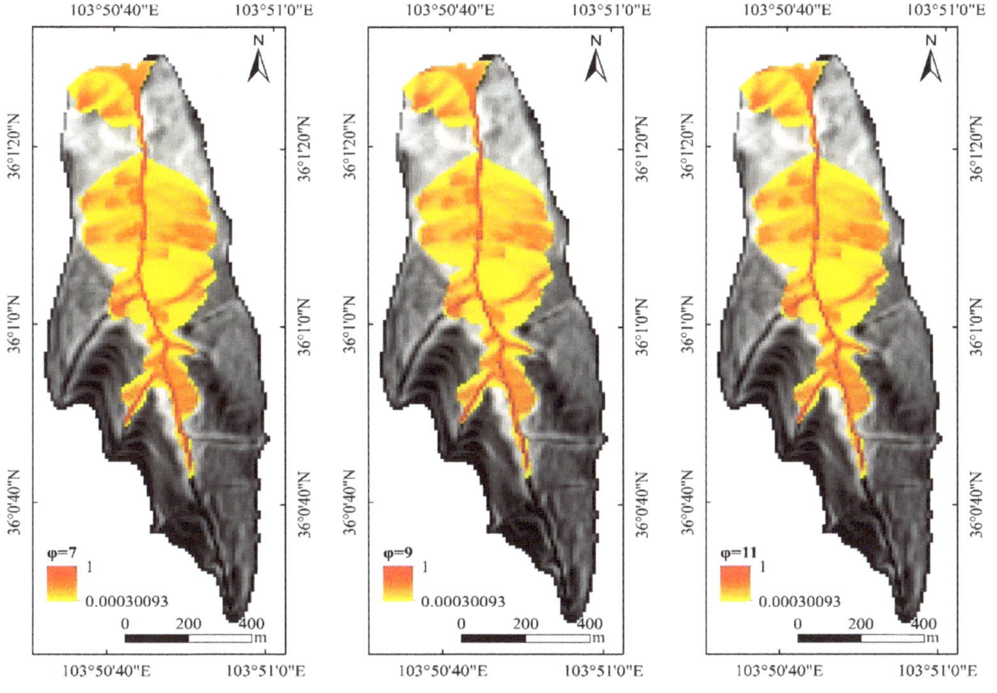

(**a**) φ = 7° (**b**) φ = 9° (**c**) φ = 11°

Figure 19. Spreading results of debris flow under different arrival angles.

4. Conclusions

Based on the acquired data of landslides in the Laolang gully, a 1:10,000 topographic map of the study area was digitized and 19 representative landslides were selected for this study. The thicknesses of the landslides were tested using a standard penetrometer. The accuracy of the measurement results were verified against the high-density electrical method. At last, a numerical model was established to simulate the occurrence and development of mudslides under different working conditions, and the propagation range was evaluated based on the propagation algorithm and the friction law. The simulation results show that:

(1) A too large or too small DEM data resolution will cause the misjudgment of debris flow source area, and the diffusion range of debris flow is not very reasonable. The 10 m resolution DEM data were considered to be good simulation data in the region, and the simulated results matched well with the potential disaster events observed in the field, which enables a reasonable evaluation of the risk of mudslides.

(2) By comparing the simulation results with the actual loose accumulations in the study area, the reasonable topographic conditions for debris flow occurrence were found. It was observed that a lower critical slope threshold allows more loose material to flow, leading to a wider spread and greater distance of the landslide.

(3) By changing the angle of arrival parameter in the SFLM model, we simulated the difference in the landslide movement process under different working conditions. The greater the angle of arrival, the more energy is lost due to friction, resulting in a shorter landslide movement distance.

Author Contributions: Conceptualization, G.B. and N.A.; methodology, G.B. and B.W.; software, N.A.; validation, G.B., N.A. and B.W.; formal analysis, G.B. and B.W.; investigation, G.B. and N.A.; resources, G.B.; data curation, B.W. and N.A.; writing—original draft preparation, G.B. and N.A.; writing—review and editing, N.A.; visualization, G.B. and N.A.; supervision, G.B.; funding acquisition, G.B. All authors have read and agreed to the published version of the manuscript.

Funding: This research was partially supported by the Special Fund for Geological Disaster Prevention and Control of Department of Natural Resources of Gansu Province, grant number 20230209GYY; partially supported by 2021 Innovation Fund Project of Gansu Provincial Bureau of Geology and Mineral Resources, grant number 2021CX13.

Data Availability Statement: The date can be obtained by contacting the corresponding author upon request.

Acknowledgments: The authors would like to thank the editor and reviewers for carefully dealing with this paper.

Conflicts of Interest: The authors declare no conflicts of interest.

References

1. Wahab, M.K.A.; Zainol, M.R.R.M.; Ikhsan, J.; Zawawi, M.H.; Abas, M.A.; Noor, N.M.; Razak, N.A.; Sholichin, M. Assessment of Debris Flow Impact Based on Experimental Analysis along a Deposition Area. *Sustainability* **2023**, *15*, 13132. [CrossRef]
2. Bai, G.; Hou, Y.; Wan, B.; An, N.; Yan, Y.; Tang, Z.; Yan, M.; Zhang, Y.; Sun, D. Performance Evaluation and Engineering Verification of Machine Learning Based Prediction Models for Slope Stability. *Appl. Sci.* **2022**, *12*, 7890. [CrossRef]
3. Zhou, Y.; Yue, D.; Liang, G.; Li, S.; Zhao, Y.; Chao, Z.; Meng, X. Risk Assessment of Debris Flow in a Mountain-Basin Area, Western China. *Remote Sens.* **2022**, *14*, 2942. [CrossRef]
4. Cui, P.; Liu, S.; Tan, W. Progress of debris flow in China. *J. Nat. Disaster* **2000**, *9*, 10–15.
5. Sha, S.; Dyson, A.P.; Kefayati, G.; Tolooiyan, A. Mathematical modelling of debris flow-boulder-barrier interactions using the Coupled Eulerian Lagrangian method. *Appl. Math. Model.* **2023**, *127*, 143–171. [CrossRef]
6. Lyu, B.; Li, Y.; Hu, P. Effects of bed sediment conditions on debris flow propagation from the two-phase flow modelling perspective. *Adv. Water Resour.* **2023**, *183*, 104592. [CrossRef]
7. Ye, Z.; Xu, Z.; Li, B.; Su, X.; Shi, G.; Meng, J.; Tian, L. Generation and origin of persistent high excess pore water pressure in debris flows. *Geomorphology* **2024**, *444*, 108954. [CrossRef]
8. Chen, M.; Tang, C.; Xiong, J.; Chang, M.; Li, N. Spatio-temporal mapping and long-term evolution of debris flow activity after a high magnitude earthquake. *Catena* **2024**, *236*, 107716. [CrossRef]
9. Rajabian, A.; Shukla, S.K. Stability Analysis of Anchor-Reinforced Soil Slopes with Taylor's Stability Chart. *Int. J. Geomech.* **2023**, *23*, 04022278. [CrossRef]
10. Dong, L.; Sun, D.; Li, X.; Zhou, Z. Interval Non-Probabilistic Reliability of a Surrounding Jointed Rockmass in Underground Engineering: A Case Study. *IEEE Access* **2017**, *5*, 18804–18817. [CrossRef]
11. Dong, L.; Sun, D.; Li, X.; Ma, J.; Zhang, L.; Tong, X. Interval non-probabilistic reliability of surrounding jointed rockmass considering microseismic loads in mining tunnels. *Tunn. Undergr. Space Technol.* **2018**, *81*, 326–335. [CrossRef]
12. Dong, L.; Sun, D.; Li, X. Theoretical and Case Studies of Interval Nonprobabilistic Reliability for Tailing Dam Stability. *Geofluids* **2017**, *2017*, 8745894. [CrossRef]
13. Ng, C.W.; Zhang, Q.; Ni, J.; Li, Z. A new three-dimensional theoretical model for analysing the stability of vegetated slopes with different root architectures and planting patterns. *Comput. Geotech.* **2021**, *130*, 103912. [CrossRef]
14. Ni, J.J.; Leung, A.K.; Ng, C.W.W.; Shao, W. Modelling hydro-mechanical reinforcements of plants to slope stability. *Comput. Geotech.* **2018**, *95*, 99–109. [CrossRef]
15. Ng, C.W.; Qu, C.; Ni, J.; Guo, H. Three-dimensional reliability analysis of unsaturated soil slope considering permeability rotated anisotropy random fields. *Comput. Geotech.* **2022**, *151*, 104944. [CrossRef]
16. Subramanian, S.S.; Ishikawa, T.; Tokoro, T. Stability assessment approach for soil slopes in seasonal cold regions. *Eng. Geol.* **2017**, *221*, 154–169. [CrossRef]
17. Alberto, G.; Remondo, J.; Cendrero, A. A methodological approach for the analysis of the temporal occurrence and triggering factors of landslides. *Geomorphology* **1999**, *30*, 95–113.
18. Thomas, C.P. Flow characteristics of large eruption-triggered debris flows at snow-clad volcanoes: Constraints for debris-flow models. *J. Volcanol. Geotherm. Res.* **1995**, *66*, 283–294.
19. Fang, Y.; Li, H.; Ai, N.; Luo, W. A study of debris flow block on the model of percolation. *Commun. Nonlinear Sci. Numer. Simulation* **1999**, *4*, 190–193. [CrossRef]
20. Gao, H.; Zhang, X. Landslide Susceptibility Assessment Considering Landslide Volume: A Case Study of Yangou Watershed on the Loess Plateau (China). *Appl. Sci.* **2022**, *12*, 4381. [CrossRef]
21. Chen, R.; Lin, M.L.; Chen, H. Mechanism of initiation of debris flow. In *Urban Disaster Mitigation: The Role of Engineering and Technology*; Pergamon: Bergama, Turkey, 1995; pp. 231–241.

22. Kang, S.; Lee, S.; Vasu, N.N.; Park, J.; Lee, D. Development of an initiation criterion for debris flows based on local topographic properties and applicability assessment at a regional scale. *Eng. Geol.* **2017**, *230*, 64–76. [CrossRef]
23. Li, Y.; Meng, X.; Guo, P.; Dijkstra, T.; Zhao, Y.; Chen, G.; Yue, D. Constructing rainfall thresholds for debris flow initiation based on critical discharge and S-hydrograph. *Eng. Geol.* **2021**, *280*, 105962. [CrossRef]
24. Legiman, M.K.A.; Mohamad, E.T.; Hasbollah, D.Z.A.; Suparmanto, E.K.; Rathinasamy, V. Contributing factors in initiation of debris flow in Malaysia. *Phys. Chem. Earth Parts A/B/C* **2023**, *129*, 103301. [CrossRef]
25. Zhou, J.; Cui, P.; Yang, X. Dynamic process analysis for the initiation and movement of the Donghekou landslide-debris flow triggered by the Wenchuan earthquake. *J. Asian Earth Sci.* **2013**, *76*, 70–84. [CrossRef]
26. Chang, M.; Liu, Y.; Zhou, C.; Che, H. Hazard assessment of a catastrophic mine waste debris flow of Hou Gully, Shimian, China. *Eng. Geol.* **2020**, *275*, 105733. [CrossRef]
27. Cui, W.; Chen, J.; Chen, X.; Tang, J.; Jin, K. Debris flow characteristics of the compound channels with vegetated floodplains. *Sci. Total Environ.* **2023**, *868*, 161586. [CrossRef]
28. Dai, F.; Li, C.; Ngai, Y. Landslide risk assessment and management: An overview. *Eng. Geol.* **2002**, *64*, 65–87. [CrossRef]
29. Huang, J.; Huang, R.; Ju, N.; Xu, Q.; He, C. 3D WebGIS-based platform for debris flow early warning: A case study. *Eng. Geol.* **2015**, *197*, 57–66. [CrossRef]
30. Sättele, M.; Bründl, M.; Straub, D. Reliability and effectiveness of early warning systems for natural hazards: Concept and application to debris flow warning. *Reliab. Eng. Syst. Saf.* **2015**, *142*, 192–202. [CrossRef]
31. Mohanty, A.; Hussain, M.; Mishra, M.; Kattel, D.B.; Pal, I. Exploring community resilience and early warning solution for flash floods, debris flow and landslides in conflict prone villages of Badakhshan, Afghanistan. *Int. J. Disaster Risk Reduct.* **2019**, *33*, 5–15. [CrossRef]
32. Zhao, Y.; Meng, X.; Qi, T.; Li, Y.; Chen, G.; Yue, D.; Qing, F. AI-based rainfall prediction model for debris flows. *Eng. Geol.* **2022**, *296*, 106456. [CrossRef]
33. Chen, J.; Luo, Y.; Xiong, J.; Zhang, S.; Xia, M.; Yang, H.; Ge, Q. A thermal-effect-based monitoring method for debris flow warning. *Geomorphology* **2022**, *400*, 108097. [CrossRef]
34. Xia, X.; Jarsve, K.T.; Dijkstra, T.; Liang, Q.; Meng, X.; Chen, G. An integrated hydrodynamic model for runoff-generated debris flows with novel formulation of bed erosion and deposition. *Eng. Geol.* **2023**, *326*, 107310. [CrossRef]
35. Timilsina, S.; Niemann, J.D.; Rathburn, S.L.; Rengers, F.K.; Nelson, P.A. Modeling Hydrologic Processes Associated with Soil Saturation and Debris Flow Initiation During the September 2013 Storm, Colorado Front Range. *Landslides* **2021**, *18*, 1741–1759. [CrossRef]
36. Chatra, A.S.; Dodagoudar, G.R.; Maji, V.B. Numerical modelling of rainfall effects on the stability of soil slopes. *Int. J. Geotech. Eng.* **2017**, *13*, 425–437. [CrossRef]
37. Lourenço, S.D.N.; Wang, G.; Kamai, T. Processes in model slopes made of mixtures of wettable and water repellent sand: Implications for the initiation of debris flows in dry slopes. *Eng. Geol.* **2015**, *196*, 47–58. [CrossRef]
38. Dong, L.; Shu, W.; Sun, D.; Li, X.; Zhang, L. Pre-Alarm System Based on Real-Time Monitoring and Numerical Simulation Using Internet of Things and Cloud Computing for Tailings Dam in Mines. *IEEE Access* **2017**, *5*, 21080–21089. [CrossRef]
39. Hu, W.; Xu, Q.; Wang, G.H.; van Asch, T.W.J.; Hicher, P.Y. Sensitivity of the initiation of debris flow to initial soil moisture. *Landslides* **2014**, *12*, 1139–1145. [CrossRef]
40. He, N.; Qu, X.; Yang, Z.; Xu, L.; Gurkalo, F. Disaster Mechanism and Evolution Characteristics of Landslide—Debris-Flow Geohazard Chain Due to Strong Earthquake—A Case Study of Niumian Gully. *Water* **2023**, *15*, 1218. [CrossRef]
41. Niu, X.; Hou, K.; Sun, H. Study on the Prevention and Control of Downhole Debris Flows Based on Disaster Chain Theory. *Water* **2023**, *15*, 2367. [CrossRef]
42. Guo, F.; Song, X.; Xie, Y.; Meng, X. A discussion on the geological hazards meteorological warning system in Gansu province. *Chin. J. Geol. Hazard Control* **2015**, *26*, 127–133.

water

MDPI

Article

Hydrogeochemical Characteristics and Formation Processes of Ordovician Limestone Groundwater in Zhuozishan Coalfield, Northwest China

Shidong Wang [1,2], Tiantian Wang [1,2,*], Zhibin Yang [1,2], Hongwei Tang [1,2], Hanjiang Lv [1,2], Feng Xu [1,2], Kaipeng Zhu [1,2] and Ziyuan Liu [1,2]

1 Technology & Engineering, Xi'an Research Institute of China Coal (Group), Corp, Xi'an 710077, China
2 Key Laboratory of Coal Mine Water Hazard Prevention and Control Technology in Shaanxi Province, Xi'an 710054, China
* Correspondence: wangthpuedu@126.com

Abstract: A comprehensive understanding of the characteristics and formation mechanisms of groundwater in mining areas is essential for the effective prevention of coal mine water and the rational management of groundwater resources. The objective of this study was to examine the hydrogeochemical characteristics and evolution of Ordovician groundwater in the Zhuozishan coal mine, located in the northwest region of China. A total of 34 groundwater samples were collected for hydrogeochemical analyses and the investigation of their evolution processes, with the aid of a piper trilinear diagram, a Gibbs diagram, and an ion ratio diagram. The results indicate that the concentration of sodium (Na^+), potassium (K^+), bicarbonate (HCO_3^-), chloride (Cl^-), sulphate (SO_4^{2-}), total dissolved solids (TDS), and pH increases from the recharge area to the discharge area, whereas the concentration of calcium (Ca^{2+}) and magnesium (Mg^{2+}) decreases. The hydrogeochemical characteristics of the runoff from Zhuozishan to Gongdeer coalfield and further southward display a notable north–south directional change. The groundwater process is primarily controlled by rock weathering action and cation exchange, with Na^+ and K^+ deriving primarily from cation exchange and only to a minor extent from halite dissolution. In conclusion, the northern part of the coalfield is characterised by a geological structure that creates a retention area with groundwater, resulting in an unordered runoff process with a complex formation mechanism. The middle region is devoid of geological constraints that would alter the flow direction, thus simplifying the process of groundwater formation. In contrast, the southern area experiences an increase in strata depth and fault blocking, which creates a retention zone, thereby rendering the groundwater formation process more complex. This research contributes to the effective management of groundwater resources in this coalfield and other mining sites.

Keywords: coal mining; groundwater hydrogeochemical characteristics; hydrogeochemical; Zhuoizishan coalfield

Citation: Wang, S.; Wang, T.; Yang, Z.; Tang, H.; Lv, H.; Xu, F.; Zhu, K.; Liu, Z. Hydrogeochemical Characteristics and Formation Processes of Ordovician Limestone Groundwater in Zhuozishan Coalfield, Northwest China. *Water* **2024**, *16*, 2398. https://doi.org/10.3390/w16172398

Academic Editors: Maurizio Barbieri, Helong Gu, Xueyi Shang and Huatao Zhao

Received: 14 July 2024
Revised: 7 August 2024
Accepted: 22 August 2024
Published: 26 August 2024

1. Introduction

In recent years, the global demand for fresh groundwater has escalated significantly due to rapid population growth, industrial expansion, and the intensification of agricultural irrigation activities [1,2]. This heightened demand has put immense pressure on groundwater resources, particularly in regions where these resources are already scarce or vulnerable to depletion [3–5]. These issues arise from a combination of natural processes and anthropogenic activities, particularly those related to extensive groundwater extraction and mining operations.

In the context of coal mining, large volumes of groundwater are often discharged from mine shafts and pits, leading to a significant decline in local groundwater levels [6–8]. This discharge can also induce hydraulic connectivity between different aquifers, thereby facilitating the migration of contaminants and resulting in widespread water quality deterioration [9]. The impact of mining activities on groundwater is thus profound, affecting both the quantity and quality of this vital resource. Understanding the geochemical characteristics and processes governing groundwater in mining regions is therefore crucial for assessing the long-term sustainability of groundwater resources and for developing effective management strategies [10].

China, with its vast coal reserves, stands as the world's leading producer and consumer of coal [5,9]. In 2019, China's coal consumption reached a staggering 2.22 billion tons, accounting for 57.7% of the country's total energy consumption [10]. However, the hydrogeological conditions underlying many of China's coal mines are complex and challenging, making these areas particularly susceptible to groundwater-related hazards, including mine water inrush incidents. From 2000 to 2014, China reported over 4500 fatalities linked to mine water inrush events, underscoring the severity of the issue [5,9]. For an extended period, techniques such as hydrochemical analysis, water quality type comparison, and multivariate statistical analysis can be effectively employed to ascertain the nature of the water inrush source at the outset of the water inrush [11,12]. This method is very effective in avoiding mine water inrush disasters.

On the one hand, the Zhuozishan coalfield, located in the westernmost region of the Ordos Basin, is one such area where the hydrogeological challenges are pronounced. The coal seams in this region are primarily located within carboniferous and Permian formations, with the Ordovician limestone aquifer posing the greatest hazard due to its potential to contribute to mine water gushing-out events. A notable incident occurred in 2010 at the Luotuoshan mine within the Zhuozishan coalfield, where a catastrophic water inrush event led to a peak inflow of 60,036 m^3/h metres per hour, causing extensive economic losses and endangering human lives. This event highlighted the critical need for a thorough understanding of the hydrochemical characteristics of groundwater in the region, as such knowledge is essential for identifying potential mine water sources and implementing preventive measures against water inrushes.

On the other hand, the Zhuozishan coalfield, which falls under the jurisdiction of the Wuhai administrative region, is situated in an arid to semi-arid climatic zone. These climatic conditions contribute to the scarcity of surface water, making groundwater the most crucial resource for meeting the drinking, domestic, and industrial water demands of Wuhai city. Notably, groundwater accounts for 42.4% of the municipal water supply, with 8% specifically sourced from the drainage of Ordovician limestone in the Zhuozishan coalfield [12]. As coal mining activities in the Zhuozishan coalfield continue to intensify, the volume of groundwater drainage from the Ordovician limestone is expected to increase correspondingly. This trend underscores the growing importance of this groundwater source for the future economic development of Wuhai city. Therefore, research on the hydrochemical characteristics and formation mechanism of the Ordovician limestone aquifer in the Zhuozishan coalfield can help scientific planning and the effective use and protection of groundwater resources in the coalfield area.

2. Material and Method

2.1. Study Area

The Zhuozishan coal mine represents a significant coal mining district within the Ordos Basin. The mine is situated between the Zhuozishan and Gangdeer coalfield, spanning a longitudinal range of 107°2′15″–107°5′45″ and a latitudinal range of 39°21′15″–39°22′41″. The Zhuozishan coalfield is currently composed of ten distinct mining areas, each bearing the name of a local village: Kabuqi, Pinggou, Luotuoshan, Baiyinwusu, Dilibangwusu, Tianyu, Hairong, Zhengfeng, Limin, and Qipanjing.

The region is characterised by aridity and semi-aridity, with significant temperature fluctuations. The annual average temperature is 7.6 °C, with a daily maximum and minimum air temperature of 36.7 °C and −31.6 °C, respectively [13]. The annual precipitation is 140 mm, while the annual evaporation is 3600 mm per year [14]. Evaporation is therefore approximately 25 times greater than precipitation. Furthermore, the area has limited surface water resources, which highlights the importance of groundwater in this region.

2.2. Hydrogeological Condition

The Zhuozishan coalfield is located in Wuhai City, Inner Mongolia Autonomous Region, and the mining area is mainly distributed between Zhuozishan and Gangdel Mountain and the southern part of Zhuozishan, with the main mining of Carboniferous and Permian coal seams (Figure 1b). The Zhuozishan coalfield is an independent karst groundwater system belonging to the Cambrian–Ordovician carbonate karst groundwater system in the Ordos Basin, and it is divided into four independent water circulation systems according to the characteristics of local hydraulic connection, water volume distribution, and replenishment conditions: the Qianli Mountain karst water subsystem (I), the Gondel Mountain karst water subsystem (II), the Zhuozishan north section karst water subsystem (III), the Zhuozishan south section karst water subsystem (II), and the southern section of the Zhuozishan karst water subsystem (IV) (Figure 1a).

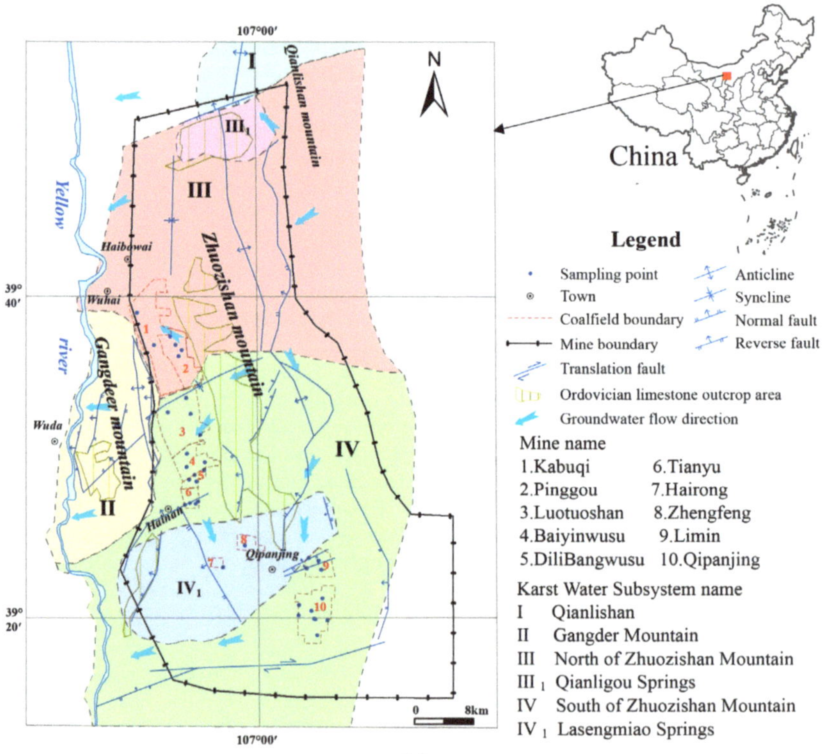

(a)

Figure 1. *Cont.*

Sytem	Symbol	Column	Thickness(m)	Lithology
Quaternary	Q		0-29.73	Dominated by a gravel layer and sandy soils
Paleogene	E		31.10-235.11	Sandy conglomerate with interbedded sandy mudstone clasts
Permian	P_2s		109.26-156. 96	Coarse sandstone in the upper part, sandy mudstone in the middle, medium and coarse sandstone in the lower part
	P_1x	No.3	17. 42-44.44	Coarse and medium fine sandstone interbedded with coloured sandy mud;
	P_1s	No.5 No.9 No.16	33.35-78.59	Sandy mudstone and coarse and medium sandstone in the upper part, medium and fine sandstone in the middle, and fine sandstone and sandy mudstone in the lower part, interbedded with coal seams
Carboniferous	C_2t		12 92-64.88	Dark grey sandy mudstone and mudstone in the upper part and greyish-white, light grey medium and fine grained sandstone in the lower part, interbedded with coal seams
Ordovician	O_1s		Thicker	Sandy mudstone, mudstone, sandstone, mudstone, sandstone, greyish-white calcareous quartz sandstone

(**b**)

Figure 1. Overview of Ordovician limestone groundwater in Zhuozishan coalfield (**a**) and composite stratigraphic column of the study area (**b**).

The coalfield is located in the northwestern part of China, where surface rivers are sparse and have low flow rates. As a result, atmospheric precipitation becomes the primary source of recharge for the Ordovician aquifer. The water level elevation of the Ordovician limestone aquifer is 1009.40–1024.47 m. Precipitation primarily recharges the aquifer in the exposed limestone area of the Zhuozishan region, from where it flows westward, crossing the Kabuchi syncline before reaching the eastern edge of the Gangdeleshan mountain range (Figure 2b1,b2). The flow is obstructed by water-blocking faults, causing the flow direction to change to north–south within the syncline. Eventually, the groundwater either discharges as springs or flows laterally towards the Yellow River, effectively dividing the Zhuozishan into two karst water subsystems, north and south, by the Kabuchi syncline.

On the other hand, groundwater also flows southward from Zhuozishan. However, due to the increasing depth of strata in the south and the obstruction caused by the Zhengyiguan strike-slip fault, a retention zone is formed, leading the groundwater to eventually discharge as springs or flow laterally towards the Yellow River (Figure 2a). Due to the limited recharge of Ordovician limestone groundwater, the groundwater circulation is relatively slow. Only in areas with karst fault structures and deeply incised karst valleys do strong runoff zones form.

2.3. Water Sampling and Analysis

A total of 34 Ordovician limestone groundwater samples were randomly collected from the 10 mining areas within the Zhuozishan coalfield. The precise locations from which samples were taken are illustrated in Figure 1. The sampling and preservation methods and processes adhered to the national standard [15]. Prior to the collection of water samples, the bottles were rinsed three times. At each sampling location, two 1 L plastic bottles were filled. One sample was utilised for cation and anion analyses, while the other was employed for in situ testing. The TDS and pH values were immediately measured in the field using a portable conductivity and pH meter (Shanghai Sanxin Company, Shanghai, China). Subsequently, the sampling bottles were labelled, sealed, and transported to the laboratory of coal mine water hazard prevention and control technology in Shaanxi Province for cation and anion determinations within 24 h. In the laboratory, all water samples were filtered through 0.45 μm millipore membrane filters prior to testing. The concentrations of anions (chloride, sulphate) were determined by anion chromatography with anion columns (Thermo Company, Waltham, MA, USA), while cations (calcium, magnesium, sodium, potassium) were measured by cation chromatography with cation columns. Furthermore, the analysis of HCO3 was conducted using the titrimetric method. The accuracy of the analysis was verified through the use of a water reference standard (NIST 1643b). In the majority of cases, the precision obtained was better than 5% RSD, with comparable accuracy.

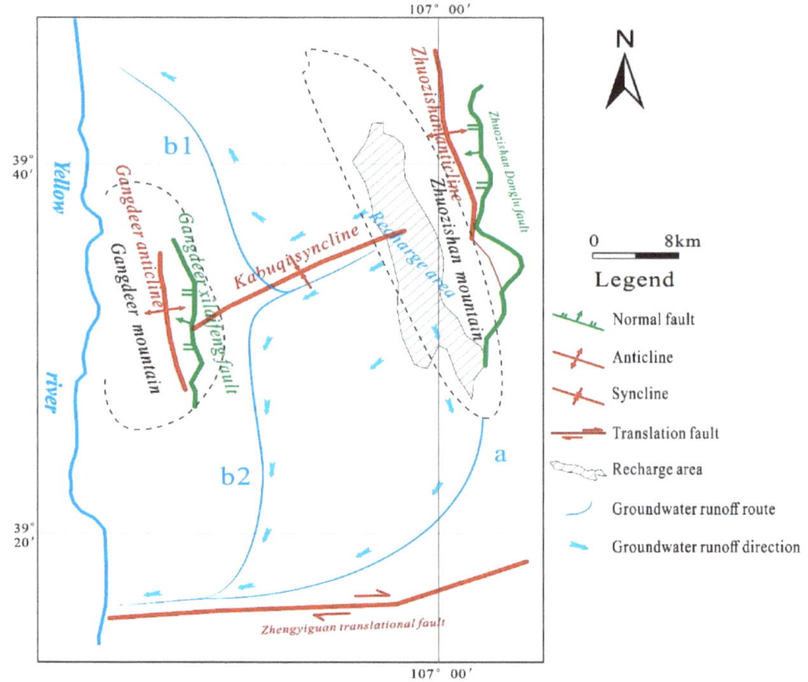

Figure 2. Recharge runoff and discharge conditions of Ordovician limestone groundwater.

3. Results

3.1. General Hydrochemistry

The principal cations and anions present in the sampled water sources are presented in Table 1. As illustrated in the table, the pH of all groundwater samples exhibited a range from 7.01 (neutral) to 7.78 (slightly alkaline). The total dissolved solids (TDSs) of the groundwater samples exhibited a range of 464.75 to 3366.52 milligrammes per litre (mg/L). The predominant cations in the limestone groundwater were sodium (Na^+) and calcium (Ca^{2+}). The concentration of sodium ions (Na^+) was found to vary from 35.19 to 900.40 mg/L. The concentration of calcium ranged from 27.61 to 603.51 mg/L (with an average of 61.21 mg/L), while that of magnesium varied from 4.5 to 133.46 mg/L.

Table 1. The results of water sample testing.

Number	Mine	K^+	Na^+	Ca^{2+}	Mg^{2+}	HCO_3^-	$SO_4{}^{2-}$	Cl^-	pH	EC (μS/cm)
1	Kabuqi	7.77	176.57	37.5	26.53	277.78	169.58	137.76	7.75	1303.13
2		10.87	160	102.35	39.8	402.05	27.58	255.1	8.4	1604.79
3	Pinggou	1.5	230	59.17	18.63	188.45	330.93	123.78	7.85	1555.77
4		5.95	147	157.78	61.61	255.72	209.92	353.66	7.35	1914.817
5		4.28	250	90.72	54.45	258.62	249.43	353.66	7.57	2005.55
6		7.68	149	123.44	69.17	274.13	389.37	195.94	8.18	1945.20
7	Luotuoshan	32	489	52.86	63.04	247.02	555.66	503.97	8.1	3062.5
8		3	61.25	67.06	35.34	218.03	117.55	88.42	7.4	937.5
9		6.4	35.19	95.4	51.97	126.09	118.13	88.42	8.45	798.44
10		3.55	68	59.38	34.59	240	109.07	86.84	8.25	971
11	Baiyinwusu	2.2	197	603.51	23.4	32.47	213.21	1237.81	8.8	3649.08
12		6.2	288	35.5	23.4	211.65	190.98	282.93	7.77	1644.06
13		3.4	58.9	27.61	47.28	55.67	126.77	114.94	8.35	726.17
14	Dilibangwusu	2.04	76.65	65.99	32.29	265.03	108.86	82.99	8.15	1005.48
15		2.56	63.5	67.86	22.31	270.3	80.65	62.17	7.01	907.81
16		3.08	63.47	58.14	30.07	275.72	90.97	68.83	7.66	933.36
17	Tianyu	2.94	74.2	73.59	26.28	264	102.6	79.5	7.48	989.06
18		1.49	63.52	54.88	28.12	232.24	88.1	72.21	8.22	871.11
19		4.32	80	59.6	26.08	267.2	97.75	76.02	7.76	971.88
20		0.66	101.43	99.43	44.23	296.56	136.86	99.37	7.3	1325.77
21	Hairong	2.32	210.05	77.51	40.24	271.29	217.34	205.8	7.77	1664.06
22	Zhengfeng	3.84	112.2	67.5	34.71	287	200.9	62.5	7.76	1221.88
23	Limin	29.8	230	63.11	42.51	249.92	293.88	247.56	7.95	1849.98
24		2	254	27.61	25.79	110.17	229.67	176.83	8.09	1440.91
25		3.2	425	59.17	40.12	226.15	470.87	380.18	8.9	2568.39
26		3.6	150.02	67.07	40.12	242.38	156.41	194.51	7.86	1363.06
27	Qipanjing	24.89	583	134.11	68.77	206.43	569.65	804.58	7.64	3773.97
28		40	633.66	145.95	56.83	223.83	680.79	813.42	7.7	4079.67
29		4.2	504	98.61	78.33	240.64	530.96	627.75	8.17	3300.16
30		6.15	580	138.06	44.89	252.82	719.48	583.54	8.45	3685.09
31		5.6	680	94.67	85.49	142.07	694.78	822.26	8.05	4041.53
32		84.06	630	82.83	75.94	131.05	629.75	751.53	7.93	3770.39
33		81.48	900.4	69.42	88.36	188.45	953.27	1069.82	7.76	5260.19
34		3.5	604	114.39	121.31	212.23	617.4	884.15	7.94	4044.69

Note: Unit of cation and anion concentration is mg/L, and pH is dimensionless.

The detection results and direction of groundwater runoff were employed to generate seven conventional ion (Ca^{2+}, Mg^{2+}, Na^+, K^+, Cl^-, SO_4^{2-}, HCO_3^-) concentration, pH, and TDS contour maps utilising Suffer software (Suffer 8.0) (Figure 3). The recharge directions of Ordovician limestone water can be classified into two categories. The initial category encompasses the flow from Zhuozishan Mountain to Gongdeer Mountain, subsequently extending in a south–north direction. The second category comprises the recharge from Zhuozishan Mountain to the south. It is evident that the distribution characteristics of the aforementioned ions (Na^+ + K^+, HCO_3^-, Cl^-, SO_4^{2-}, TDS, and pH) generally increase from the recharge area to the discharge area. In contrast, the distribution characteristics of Ca^{2+} and Mg^{2+} demonstrated an inverse relationship. This phenomenon is particularly evident in the central region. The contour distribution of the various indicators is relatively dense in the northern and southern parts, indicating significant variation in the hydrochemical indicators.

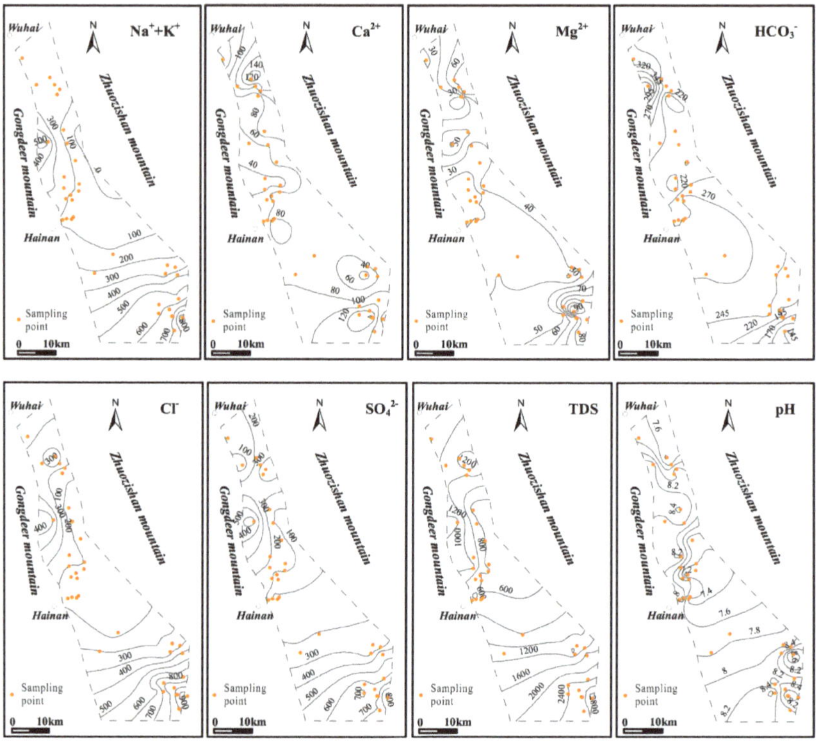

Figure 3. Contour of the concentration of ions.

3.2. Hydrochemical Types

The Piper trilinear diagram is an effective tool for identifying the relationships between diverse dissolved components and various types of groundwater, based on their hydrochemical characteristics [16,17]. As illustrated in the Piper diagram (Figure 4), the predominant hydrochemical types of Ordovician limestone water in the Zhuozishan coalfield were identified as SO_4·Cl-Ca·Na, HCO_3·SO_4·Cl-Ca·Na, HCO_3·Cl-Ca·Na, and HCO_3^- Ca·Na. The hydrochemical types of groundwater can be classified according to the direction of groundwater runoff, resulting in two distinct categories, as illustrated in the figure. The hydrochemical type undergoes a transformation from SO_4·Cl-Ca·Na in the upstream region to HCO_3·SO_4·Cl-Ca·Na, HCO_3·Cl-Ca·Na, or HCO_3-Ca·Na as the groundwater flows from the Zhuozishan Mountain to the Gongdeer Mountain and then in a north–south direction. Secondly, the runoff from the Zhuozishan Mountain to the south consistently exhibits the hydrochemical type SO_4·Cl-Ca·Na. As illustrated in the TDS contour map, the TDS concentration of the runoff originating from the Zhuozishan Mountain and traversing the Gongdeer Mountain before continuing in a northerly direction ranges between 1200 and 1400 mg/L. In contrast, the concentration of the runoff directed southwards falls within the 600–1000 mg/L range. The concentration of total dissolved solids (TDSs) in the runoff from the Zhuozishan Mountain to the south is 1200–2800 mg/L, exhibiting a gradual increase. It is evident that the proportion of HCO_3^- (mg equivalent) increases, while the TDS concentration undergoes only slight fluctuations throughout the runoff process from the Zhuozishan Mountain to the Gongdeer Mountain and subsequently to the south or north. During the runoff process from the Zhuozishan Mountain to the south, the hydrochemical type remains unaltered, yet the TDS concentration rises gradually, suggesting that hydrochemical reactions occur within the groundwater during this process.

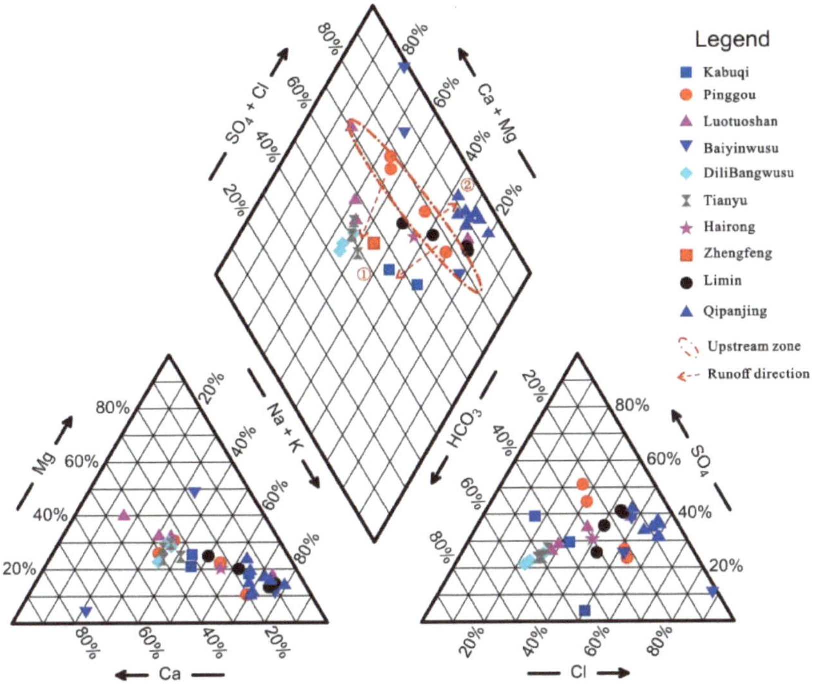

Figure 4. Piper trilinear diagram of Ordovician limestone water in Zhuozishan coalfield. (① The primary control action in this case is rock weathering; ② The control action of groundwater shifts from rock dominance to evaporation domi-nance).

4. Discussion

4.1. Rock Weathering Action

The Gibbs diagram [18] constituted an invaluable tool for the initial investigation of the processes involved in the formation of surface water, and it has since become a widely employed methodology in groundwater research [19]. The principal control factors can be classified into three categories: those pertaining to evaporation and concentration, those associated with rock weathering, and those related to precipitation (Figure 5). The analytical results demonstrate that the groundwater hydrochemical process can be divided into two categories. The first category comprises runoff from Zhuozishan Mountain, which flows north–south. The primary control action in this case is rock weathering. The second category encompasses runoff from Zhuozishan Mountain to the south. In this instance, the control action of groundwater shifts from rock dominance to evaporation dominance, progressing from the upstream to the downstream. However, the Ordovician limestone groundwater is located at a considerable depth, and thus will not be subjected to evaporation and concentration. It can thus be postulated that a reverse cation exchange effect may occur in the runoff direction from Zhuozishan Mountain to the south, resulting in an increase in the concentration of certain ions.

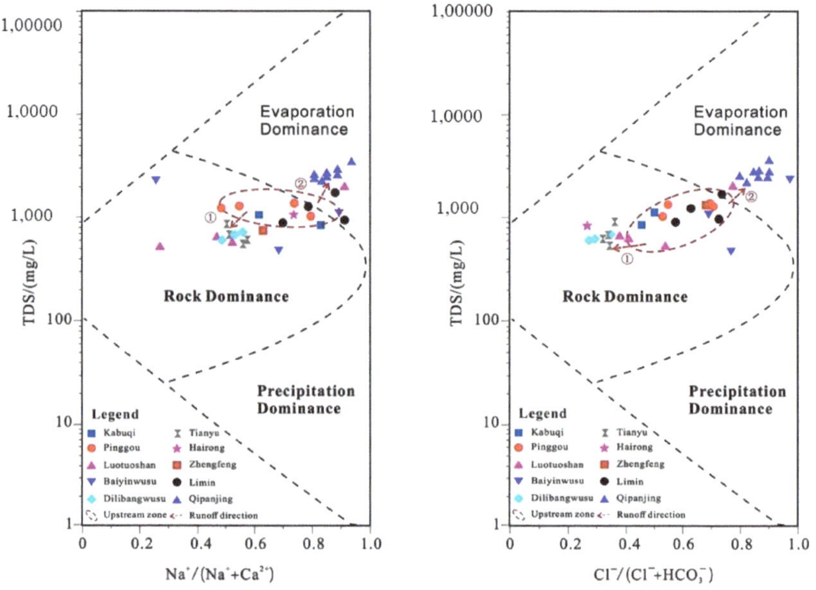

Figure 5. Gibbs diagram of Ordovician limestone water. (① The primary control action in this case is rock weathering; ② The control action of groundwater shifts from rock dominance to evaporation domi-nance).

Cation exchange is also a vital process affecting groundwater chemical features [20–22]. Typically, two chloro-alkaline indices (*CAI*-1 and *CAI*-2) proposed by Schoeller (1965) [23] are important ways of researching the occurrence of cation exchange, where the units of all ions are meq/L. When *CAI*-1 and *CAI*-2 are negative, this indicates that an ion exchange effect occurs, in which Na^+ and (or) K^+ in water-bearing media are replaced by Ca^{2+} and (or) Mg^{2+} in groundwater. But when *CAI*-1 and *CAI*-2 are positive, the reverse cation exchange effect occurs [24]. The calculation equations of *CAI*-1 and *CAI*-2 can be estimated as follows [25]:

$$CAI-1 = (Cl^- - (Na^+ + K^+))/Cl^- \tag{1}$$

$$CAI-2 = (Cl^- - (Na^+ + K^+))/\left(HCO_3^- + SO_4^{2-} + CO_3^{2-} + NO_3^-\right) \tag{2}$$

In this study, samples from the north and central parts of Zhuozishan coalfield groundwater are drawn in the lower left area, whereas samples from the south part are mostly in the upper right area of Figure 6a. This suggests that an ion exchange effect (Equation (3)) occurs in the runoff from the Zhuozishan to Gongdeer coalfield and then in a south–north direction, and reverse cation exchange (Equation (4)) occurs in the runoff from Zhuozishan coalfield to the south. In other words, groundwater cation exchanges of these two parts occur in reverse.

$$Ca^{2+}(Mg^{2+})(water) + Na^+(K^+)(rock) \rightarrow Na^+(K^+)(water) + Ca^{2+}(Mg^{2+})(rock) \tag{3}$$

$$Na^+(K^+)(water) + Ca^{2+}(Mg^{2+})(rock) \rightarrow Ca^{2+}(Mg^{2+})(water) + Na^+(K^+)(rock) \tag{4}$$

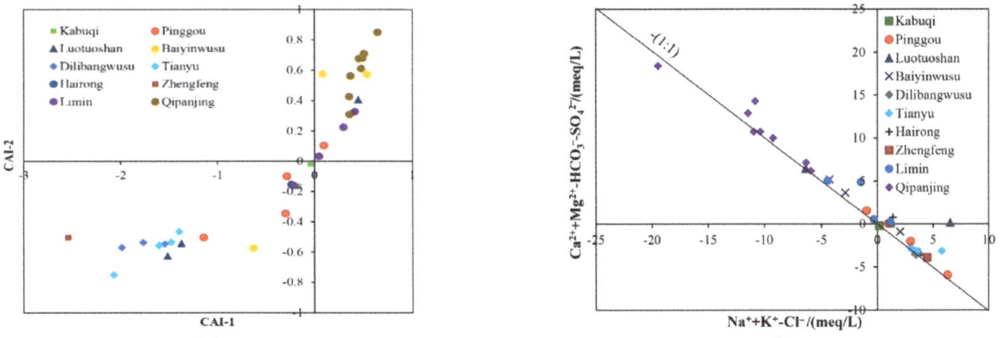

(a) (b)

Figure 6. Chloro-alkaline indices of water (**a**) and (Ca^{2+} + Mg^{2+} − HCO_3^- − SO_4^{2-})/(Na^+ + K^+ − Cl^-) (**b**).

In order to verify the hypothesis of cation exchange reactions, the relationship between (Na^+ + K^+ − Cl^-) and [(Ca^{2+} + Mg^{2+}) − (HCO_3^- − $SO_4{}^{2-}$)] was examined (Figure 6b). The values of (Na^+ + K^+ − Cl^-) indicate the extent of Na^+ gains or losses resulting from halite dissolution [12]. Furthermore, the values of [(Ca^{2+} + Mg^{2+}) − (HCO_3^- − $SO_4{}^{2-}$)] demonstrate the extent of Ca^{2+} and Mg^{2+} gains or losses resulting from the dissolution of gypsum, calcite, and dolomite [26]. Figure 6a demonstrates that all groundwater samples are distributed along or in close proximity to the line with a slope of −1, indicating that the Ordovician limestone groundwater is subject to cation exchange. The samples from coal mines situated in the northern and central regions of the Zhuozishan coalfield are predominantly distributed in the fourth quadrant of the coordinate axis, whereas those from the southern area are primarily situated in the second quadrant of the coordinate axis. Furthermore, the distance from Zhuozishan Mountain (the source of groundwater recharge) is inversely proportional to the distance from the origin of the coordinates, indicating that the strength of the cation exchange effect increases with distance from the recharge area.

4.2. Main Ion Sources

The ratio of sodium to potassium and chloride can be employed as a means of investigating the sources of sodium and potassium in groundwater. The proximity of the samples to the 1:1 aquiline indicates that the primary sources of Na^+ and K^+ are halite dissolution [27]. Figure 7a demonstrates that groundwater samples from the south and north of the Zhuozishan coalfield exhibit a significant deviation from the 1:1 line, indicating that Na^+ and K^+ originate from halite dissolution and are minimally influenced by cation exchange.

Prior research has demonstrated that if sulphate rock dissolution represents the primary process within a groundwater system, the (Ca^{2+} + Mg^{2+})/$SO_4{}^{2-}$ ratio will be approximately equal to 1 [28,29]. Figure 7b demonstrates that the majority of the water samples are distributed above the 1:1 line at a ratio of approximately 2:1, indicating that the Ca^{2+} and Mg^{2+} in the Ordovician limestone groundwater are derived from carbonate or silicate dissolution, rather than sulphate minerals. Furthermore, if carbonate dissolution is the primary process in a groundwater system, the (Ca^{2+} + Mg^{2+})/ HCO_3^- ratio will be approximately 1. Figure 7c demonstrates that only a limited number of groundwater samples plot near the 1:1 line. The water samples from the southern and northern regions of the coalfield plot above the 1:1 line, whereas those from the central region are predominantly distributed below this line. In combination with Figure 7b,c, it can be concluded that the majority of Ca^{2+} and Mg^{2+} in the groundwater system is derived from silicate dissolution, with a smaller contribution from other rock types.

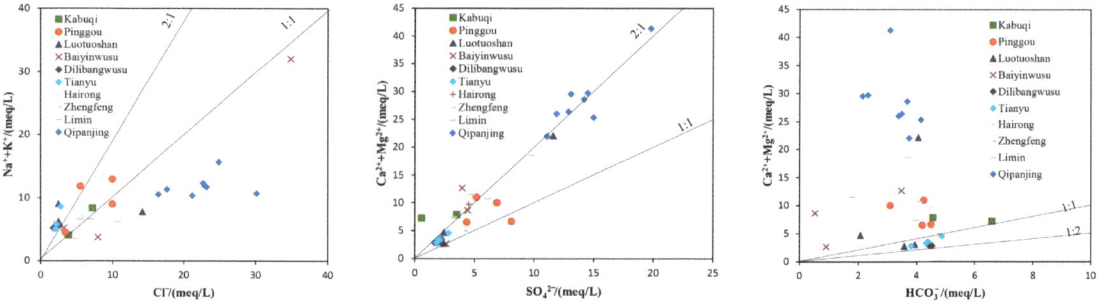

(**a**) $(Na^+ + K^+) / Cl^-$ (**b**) $(Ca^{2+} + Mg^{2+}) / SO_4^{2-}$ (**c**) $(Ca^{2+} + Mg^{2+}) / HCO_3^-$

Figure 7. Ratio relation of main ions in Ordovician limestone water.

4.3. *Mechanisms of Hydeogeochemistry*

Based on the previous sections, the Ordovician groundwater runoff in the Zhuozishan coalfield can be classified into two primary flow directions. One direction flows from Zhuozishan Mountain to Gongdeer Mountain, then continues in a north–south direction. The other direction originates from Zhuozishan Mountain and flows southward. Furthermore, hydrochemical analysis reveals that the Zhuozishan coalfield can be divided into three distinct regions: the "northern", "central", and "southern" sectors. The hydrochemical processes active in each of these regions are summarised as follows:

4.3.1. Northern Part

Following replenishment from Zhuozishan Mountain, the groundwater in the northern region flows westward. The groundwater then traverses the Kabuqi syncline, reaching the eastern flank of Gongdeer Mountain, where it encounters a water-blocking fracture. As a result, the groundwater alters its course to a north–south direction within the syncline. The sudden change in the direction of groundwater runoff causes the formation of a retention area of groundwater in the Kabuqi syncline. Furthermore, the constant recharge of groundwater results in a disruption to the direction of groundwater runoff in the retention area (Figure 8). Consequently, groundwater flows to the downstream area for an extended period, resulting in a series of interconnected and non-uniform contours of hydrochemical indices (Figure 4). This consequently leads to a more complex hydrochemical typology and process in this region.

4.3.2. Central Part

The groundwater in the central region is derived from two primary sources: runoff from the north and the recharge area of Zhuozishan Mountain. The former flows downstream through the fault zone, while the latter contributes to the groundwater flow in the southern part of the Zhuozishan coalfield. It can therefore be concluded that the groundwater in the central region has not been obstructed by geological structures and that the direction of runoff remains unchanged. The regular alteration in the contours of the hydrochemical indices (Figure 4) is characterised by a hydrochemical type dominated by HCO_3·Cl-Ca·Na (Figure 5). The hydrochemical process is primarily controlled by rock weathering (Figure 6) and cation exchange (Figure 7).

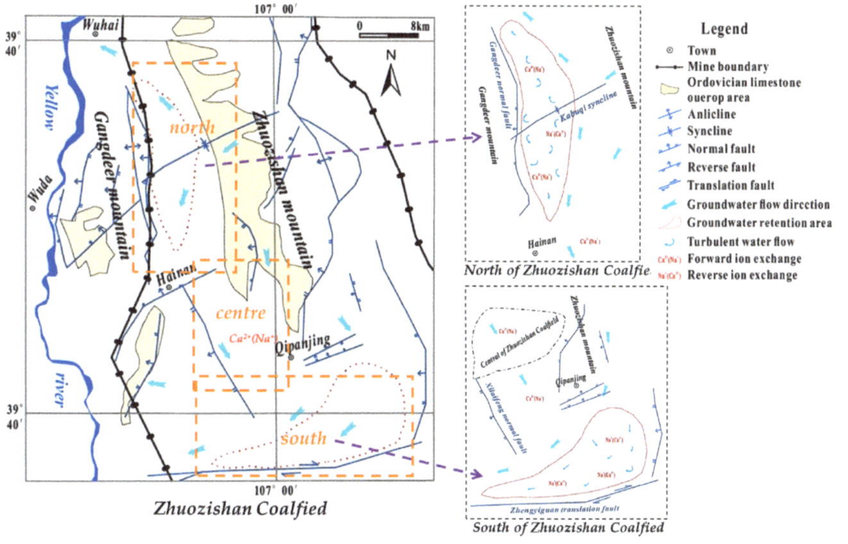

Figure 8. Schematic diagram for the formation of Ordovician groundwater in Zhuozishan coalfield.

4.3.3. Southern Part

The groundwater in the southern region is derived primarily from the runoff in the north or the recharge area of Zhuozishan Mountain. However, as a result of the gradual deepening of strata in the south and the blocking of the Zhengyiguan strike-slip fault, a retention zone has been formed (Figure 8). Subsequently, groundwater flows in a southwesterly direction. Consequently, the hydrochemical indices display a dense and regular distribution pattern, as illustrated in Figure 3. The hydrochemical type is consistently dominated by the SO_4·Cl-Ca·Na type (Figure 4), with a more prominent reverse cation exchange effect (Figure 7).

5. Conclusions

The characteristics of the Ordovician limestone groundwater in the Zhuozishan mining area, located in northwestern China, as well as its hydrochemical processes and mechanisms, were analysed through a series of groundwater samples collected from the Zhuozishan mining area. The conclusions of the study can be summarised as follows:

(1) The hydrogeochemical characteristics of the Ordovician limestone groundwater in Zhuozishan are closely related to the direction of groundwater runoff. The concentrations of Na^+, K^+, Cl^-, SO_4^{2-}, HCO_3^-, TDS, and pH increase from the recharge area to the discharge area, whereas those of Ca^{2+} and Mg^{2+} decrease. The concentration contours of Na^+, K^+, Cl^-, SO_4^{2-}, HCO_3^-, Ca^{2+}, Mg^{2+}, TDS, and pH are densely distributed in the north and south of Zhuozishan, indicating that these chemical indices show significant variations.

(2) A significant alteration in the hydrogeochemical types of the previous runoff direction (from Zhuozishan Mountain to Gongdeer Mountain and then to the south–north directional runoff) can be observed, whereas the latter is consistently dominated by SO_4·Cl-Ca·Na. This indicates different hydrochemical effects have occurred in their runoff process. Furthermore, according to the analysis of the Gibbs diagram and chloro-alkaline index, it can be concluded that the formation processes of groundwater in the two runoff directions are governed by distinct mechanisms. Specifically, the former is affected by rock weathering action and cation exchange, whereas the latter is controlled by rock weathering action to the evaporation concentration and reverse cation exchange. In terms of ion sources, Ca^{2+} and Mg^{2+} mainly originate from the silicate dissolution, while Na^+ and K^+ are mainly derived from cation exchange, with little halite dissolution.

(3) The formation of three hydrochemical characteristic areas with different flow directions has been identified in the study area. In the northern part, the groundwater is influenced by a complex geological structure, resulting in a retention area and disordered runoff direction, with intricate hydrochemical types and processes. The central zone, however, is unaffected by geological constraints and has a uniform flow direction, resulting in a simpler hydrochemical type, mainly influenced by rock weathering and cation exchange. In the southern part, deepening of the strata and fault blockage result in a hydrochemical type similar to that of the recharge area and strong reverse cation exchange.

Author Contributions: Conceptualisation, S.W. and T.W.; methodology, T.W.; software, Z.Y. and H.T.; validation, S.W. and T.W.; formal analysis, S.W., T.W. and H.L.; investigation, F.X. and K.Z.; data curation, K.Z., Z.L. and F.X.; writing—original draft preparation, S.W. and T.W.; writing—review and editing, S.W.; visualisation, H.L., H.T. and Z.Y.; supervision, S.W. and T.W. All authors have read and agreed to the published version of the manuscript.

Funding: This research was funded by a Key project of science and technology innovation of China Coal Science and Technology Group (2024-TD-ZD009, 2022-2-TD-ZD005), Science and technology innovation fund of China Coal Science and Technology Group (Grant No. 2023XAYJS13), Foundation of Shaanxi (Grant No.2023-JC-QN-0291).

Data Availability Statement: The data used to support the findings of this study are available from the corresponding author upon request.

Conflicts of Interest: Author Shidong Wang, Tiantian Wang, Zhibin Yang, Hongwei Tang, Hanjiang Lv, Feng Xu, Kaipeng Zhu, and Ziyuan Liu were employed by the company Technology & Engineering, Xi'an Research Institute of China Coal (Group), Corp. The remaining authors declare that the research was conducted in the absence of any commercial or financial relationships that could be construed as a potential conflict of interest.

References

1. Rajesh, R.; Brindha, K.; Elango, L. Groundwater quality and its hydrochemical characteristics in a shallow weathered rock aquifer of southern India. *Water Qual. Expo. Health* **2015**, *7*, 515–524. [CrossRef]
2. Wang, T.T.; Yang, J.; Li, G.Q. The hydrogeochemical characteristics and formation mechanism of high-fluoride mine water. *J. Clean. Prod.* **2023**, *430*, 139671. [CrossRef]
3. Jin, D.; Wang, T.; Zhao, B.; Li, D.; Zhou, Z.; Shang, H. Distribution characteristics and formation mechanism of high salinity groundwater in northeast Ningdong Coalfield. *Coal Geol. Explor.* **2022**, *50*, 118–127. [CrossRef]
4. Wang, T.; Jin, D.; Yang, J. Heavy metal pollution characteristics and source analysis of water drainage from a mine in Inner Mongolia. *Coal Geol. Explor.* **2021**, *49*, 45–51. [CrossRef]
5. Wang, H.; Dong, S.; Shang, H.; Wang, T.; Yang, J.; Zhao, C.; Zhnag, Q.; Zhou, Z.; Liu, J.; Hou, Y. Domestic and foreign progress of mine water treatment and resource utilization. *Coal Geol. Explor.* **2023**, *51*, 222–236. [CrossRef]
6. Wang, T.; Zhao, W.; Wang, Z.; Zhou, Z.; Yang, J.; Xu, F.; Xue, J.; Li, G. Occurrence, Main Source and Health Risks of Fluorine in Mine Water. *Expo. Health* **2024**, 1–14. [CrossRef]
7. Jin, D.; Li, C.; Liu, Y.; Cao, H.; Ren, D.; Wang, H.; Zhang, J.; Huang, Y.; Yang, G.; Guo, K.; et al. Characteristics of roof water hazard of coal seam in Huanglong Coalfield and key technologies for prevention and control. *Coal Geol. Explor.* **2023**, *51*, 205–213. [CrossRef]
8. Wang, T.; Jin, D.; Xue, J.; Ji, H.; Shang, H.; Zhou, Z.; Yang, J.; Cao, Y. Source and release law of fluorine in water-rock system of coal mine goaf in contiguous area of Shaanxi and Inner Mongolia. *Coal Geol. Explor.* **2023**, *51*, 252–262. [CrossRef]
9. Gu, H.; Lai, X.; Tao, M.; Momeni, A.; Zhang, Q. Dynamic mechanical mechanism and optimization approach of roadway surrounding coal water infusion for dynamic disaster prevention. *Measurement* **2023**, *223*, 113639. [CrossRef]
10. Zhang, C.; Luo, B.; Xu, Z.; Sun, Y.; Feng, L. Research on the Capacity of Underground Reservoirs in Coal Mines to Protect the Groundwater Resources: A Case of Zhangshuanglou Coal Mine in Xuzhou, China. *Water* **2023**, *15*, 1468. [CrossRef]
11. Sun, W.; Zhou, W.; Jiao, J. Hydrogeological Classification and Water Inrush Accidents in China's Coal Mines. *Mine Water Environ.* **2016**, *35*, 214–220. [CrossRef]
12. Wang, Y.; Shi, L.; Wang, M.; Liu, T. Hydrochemical analysis and discrimination of mine water source of the Jiaojia gold mine area, China. *Environ. Earth Sci.* **2020**, *79*, 123. [CrossRef]
13. Li, P.; Tian, R.; Liu, R. Solute geochemistry and multivariate analysis of water quality in the Guohua phos-phorite mine, Guizhou Province, China. *Expo. Health* **2018**, *11*, 81–94. [CrossRef]
14. Yang, Z.B. Prevention and Countermeasures of Floor Water Disaster in Lower-group Coalseam Mining in Zhuozishan Coalfield. *Coal Min. Technol.* **2015**, *20*, 133–136. (In Chinese) [CrossRef]

15. Wang, S.D.; Zhu, K.P.; Tang, H.W. Evolution Mechanism and Circulation Characteristics of Ordovician Lime-stone Water in Zhuozishan Coalfield. *Min. Saf. Environ. Prot.* **2017**, *44*, 116–120. [CrossRef]
16. Ministry of Environmental Protection of the PR China. *Water Quality Sampling—Technical Regulation of the Preservation and Handling of Samples*; HJ 493–2009; China Environmental Science Press: Beijing, China, 2009. (In Chinese)
17. Tiwari, A.K.; Singh, A.K.; Singh, M.P. Hydrogeochemical analysis and evaluation of surface water qual-ity of pratapgarh district, uttar pradesh, India. *Appl. Water Sci.* **2017**, *7*, 1609–1623. [CrossRef]
18. Shakya, B.M.; Nakamura, T.; Shrestha, S.D.; Nishida, K. Identifying the deep groundwater recharge processes in an intermountain basin using the hydrogeochemical and water isotope characteristics. *Hydrol. Res.* **2019**, *50*, 1216–1229. [CrossRef]
19. Gibbs, R.J. Mechanisms controlling world water chemistry. *Science* **1970**, *170*, 795–840. [CrossRef]
20. Li, P.; Li, X.; Meng, X.; Li, M.; Zhang, Y. Appraising groundwater quality and health risks from contamination in a semiarid region of northwest China. *Expo. Health* **2016**, *8*, 361–379. [CrossRef]
21. Pehlivan, R. Potability and Hydrogeochemical Characteristics of the Karasu Stream Water, Sakarya, Turkey. *Geochem. Int.* **2020**, *58*, 1075–1081. [CrossRef]
22. Zhang, B.; Zhao, D.; Zhou, P.; Qu, S.; Liao, F.; Wang, G.C. Hydrochemical Characteristics of Groundwater and Dominant Water–Rock Interactions in the Delingha Area, Qaidam Basin, Northwest China. *Water* **2020**, *12*, 836. [CrossRef]
23. Shi, Z.; Liao, F.; Wang, G.; Xu, Q.; Mu, W.; Sun, X. Hydrogeochemical Characteristics and Evolution of Hot Springs in Eastern Tibetan Plateau Geothermal Belt, Western China: Insight from Multivariate Statistical Analysis. *Geofluids* **2017**, *2017*, 6546014. [CrossRef]
24. Schoeller, H. Qualitative Evaluation of Groundwater Resources. In *Methods and Techniques of Groundwater Investigations and Development*; UNESCO: Paris, France, 1965; pp. 54–83.
25. Liu, J.; Gao, Z.; Wang, Z.; Xu, X.; Su, Q.; Wang, S.; Qu, W.; Xing, T. Hydrogeochemical processes and suitability assessment of groundwater in the Jiaodong Peninsula, China. *Environ. Monit. Assess.* **2020**, *192*, 1–17. [CrossRef]
26. Hou, G.C.; Lin, X.Y.; Su, X.S.; Wang, X.Y.; Liu, J. Groundwater System in Ordos Cretaceous Artisan Basin. *J. Jilin Univ. (Earth Sci. Ed.)* **2006**, *36*, 391–398. (In Chinese) [CrossRef]
27. Ke, X.M.; Li, Y.J.; Wang, W. Hydrogeochemical characteristics and processes of thermokarst lake and groundwater during the melting of the active layer in a permafrost region of the Qinghai-Tibet Plateau, China. *Sci. Total Environ.* **2022**, *851*, 158183. [CrossRef]
28. Li, P.; Wu, J.; Qian, H.; Zhang, Y.; Yang, N.; Jing, L.; Yu, P. Hydrogeochemical characterization of groundwater in and around a wastewater irrigated forest in the southeastern edge of the Tengger Desert, northwest China. *Expo. Health* **2016**, *8*, 331–348. [CrossRef]
29. Li, P.; Wu, J.; Rui, T.; Song, H.; He, X.; Xue, C.; Kang, Z. Geochemistry, Hydraulic Connectivity and Quality Appraisal of Multilayered Groundwater in the Hongdunzi Coal Mine, Northwest China. *Mine Water Environ.* **2018**, *37*, 1–16. [CrossRef]

water

Article

Research on the Development Height Prediction Model of Water-Conduction Fracture Zones under Conditions of Extremely Thin Coal Seam Mining

Hongsheng Wang [1,2], Jiahao Tian [1,*], Lei Li [1,2], Dengfeng Chen [1], Yuxin Yuan [1,2] and Bin Li [1]

1 College of Energy, Science and Engineering, Xi'an University of Science and Technology, Xi'an 710054, China; cumt_whs@163.com (H.W.); lileiyjs@163.com (L.L.); 21203077025@stu.xust.edu.cn (D.C.); yuanyuxin@126.com (Y.Y.); lbblibin@163.com (B.L.)

2 Institute of Rock Burst Prevention and Control, Xi'an University of Science and Technology, Xi'an 710054, China

* Correspondence: 22203077031@stu.xust.edu.cn

Abstract: Addressing the difficult problem of predicting the height of water-conducting fracture zones in shallow and thin coal seams, a prediction model of water-conduction fracture zones based on a backpropagation (BP) neural network was developed by integrating theoretical analysis, field measurements, and algorithmic advancements. Firstly, through overburden migration analysis and correlation tests, the height index system of the water-conducting fracture zone was determined. This system includes mining height, buried depth, dip angle, working face width, and overburden rock lithology, with five groups of characteristic parameters. Then, 35 pairs of minefield-measured data were collected to establish the measured height data set of the water-conducting fracture zone. Secondly, a BP neural network prediction model and a traditional support vector regression (SVR) prediction model were constructed based on a Pytorch framework, and the models were trained and tested by selecting data sets. Thirdly, the optimal prediction model was determined by comparing the model with the empirical model and multiple regression model of mining regulations for coal pillar maintenance and pressure in buildings, water bodies, railways, and main shafts. Finally, a typical mine was selected for application to verify the suitability of the optimal model. The results show that: (1) the predicted value of the neural network model is consistent with the change trend of the measured value, which accords with the theoretical law; (2) compared with traditional forecasting methods, the error of the BP neural network prediction model is stable and the prediction effect is the best; (3) dropout can effectively mitigate mitigation training overfitting, achieve regularization, and improve prediction accuracy; (4) the field application further verified that the BP neural network model is the best for predicting the height of water-conducting fracture zones of extremely thin coal seams, and the research results can provide technical guidance for similar fragile coal seams.

Keywords: extremely thin coal seam; water-conducting fracture zone; multiple regression model; BP neural network

Citation: Wang, H.; Tian, J.; Li, L.; Chen, D.; Yuan, Y.; Li, B. Research on the Development Height Prediction Model of Water-Conduction Fracture Zones under Conditions of Extremely Thin Coal Seam Mining. *Water* **2024**, *16*, 2273. https://doi.org/10.3390/w16162273

Academic Editors: Helong Gu, Xueyi Shang and Huatao Zhao

Received: 12 July 2024
Revised: 8 August 2024
Accepted: 9 August 2024
Published: 12 August 2024

1. Introduction

Coal mining disrupts the initial equilibrium state of the rock mass, causing bending deformation in the overlying strata of the coal seam. This process forms mining fractures, known as water-conducting fracture zones [1]. The formation of water-conducting fracture zones provides an efficient channel for groundwater resources but also results in the degradation and destruction of surface ecology after coal mining [2]. Therefore, in order to regulate the coal industry, the State Administration of Coal Industry formulated the Regulations on Coal Pillar Maintenance and Coal Pressure Mining in Buildings, Water Bodies, Railways, and Main Shafts and Lanes [3], which is a standard to predict the height of water-conducting fracture zones by an empirical formula. However, since the cumulative height of a single layer involved in this empirical formula ranges from 1 to 3 m, its applicability is relatively limited in the case of extremely thin coal seams. Looking at the existing literature, few scholars have conducted in-depth research on this topic, and a targeted prediction method for extremely thin coal seams is urgently needed in the field of mining engineering.

The development height of water-conducting fracture zones is influenced by numerous factors. The main challenge in predictions is to determine the degree of influence of each factor on the water-conducting fracture zone to ensure the accuracy of the height prediction. Currently, scholars both domestically and internationally have conducted studies on predicting water-conducting fracture zones through theoretical analysis, engineering analogies, numerical simulations, physical simulation analyses, and field measurements. Many scholars have conducted research and exploration using masonry beam theory [1], key layer theory of overlying rock [4–11], transfer rock beam theory [12,13], etc. They have employed physical similarity simulation methods [14–16], numerical analysis methods [17], and numerical simulation software [18–24] to simulate the development laws of mining overlying rock in order to enhance prediction accuracy. The development, law, and mechanical mechanism of water fracturing induced by mining in various overburden structures are disclosed. Compared with theoretical analysis, numerical simulation, and other methods, field measurement is the most intuitive method to study the development of overlying rock fractures. Guo et al. [25], Lin et al. [26], Zhu et al. [27], and others have observed the height of water-conducting fracture zones in different areas through borehole observations. The analysis of the field-measured data provides a robust reference index for determining the depth of the water-conducting fracture zone in the mining area. In addition, due to the advancements in computer science, machine learning methods have been successfully applied to predict the development height of water-conducting fracture zones. Many domestic scholars have started to investigate more precise prediction models. Li et al. [28] established a prediction model for water-conducting fracture zones using a backpropagation (BP) neural network. The BP neural network prediction model has gradually gained attention from many scholars. Chai et al. [29] used a genetic algorithm–support vector regression (GA-SVR) to establish a prediction model for predicting the height of water-conducting fracture zones. Lou and Tan [30] used a particle swarm optimization–BP (PSO-BP) neural network to perform global optimization predictions for the height of water-conducting fracture zones. Shi et al. [31] adopted the PCA-GA-Elman optimization model to effectively mitigate the interaction between factors, enhancing the accuracy of the prediction results. However, a relatively large amount of data was needed. Wu et al. [32] utilized the radial basis function neural network (RBFNN) method to predict the development height of water-conducting fracture zones under fully mechanized mining conditions. These machine learning methods provide a new way to predict the height of water-conducting fracture zones more accurately. The scholars mentioned above have conducted a comprehensive and in-depth analysis of the development height of water-conducting fracture zones under various overlying rock structures. This study provides crucial support for predicting the development height of water-conducting fracture zones. However, previous studies have mainly focused on the general situation, and research on extremely thin coal seams is relatively limited.

Based on previous studies, we collected drilling data from the Niuwu Mining area in Fuxian County, China, focusing on the No. 3 extremely thin coal seam background. Firstly, the primary influencing factors were identified through correlation analysis. Secondly, IBMSPSS Statistics 27 software was used to construct both a multiple linear regression model and a multiple nonlinear regression model, while Python was utilized to develop a BP neural network model. By comparing the measured and predicted values, the evolution law of water-conducting fracture zones in extremely thin coal seams was analyzed. The applicability of the optimal model was verified to enhance the prediction of the development height of water-conducting fracture zones in very thin coal seams. This study provides a reference for further research on predicting the development height of water-conducting fracture zones.

2. Project Overview

The Fuxian Niuwu Mining area is located in Niuwu Town, Fuxian County, within the administrative division of Yan'an City, Shaanxi Province, China. The regional reserves of resources are 819 Wt, mining thickness is 0.5~0.69 m, and the average thickness of coal is 0.6 m. The mining area is located in the Yan'an slope of the Ordos Depression basin and displays typical stable platform characteristics. The strata are inclined to the northwest, and the structure is monoclinal with a gentle dip angle. The columnar shape of the rock strata at the top and bottom of the working face is illustrated in Figure 1.

3. Coal roof and floor rock strata are columnar				
Strata division	Name of stratum	Bed thickness/m	Bar chart	Lithology description
Basic roof	Medium grained sandstone	3.1~5.6		Gray medium fine sandstone, mainly quartz feldspar, calcareous cement.
Immediate roof	Fine grained sandstone	1.3~2.5		It has wavy bedding and horizontal bedding.
False roof	mudstone	0.3~0.8		Gray-black mudstone, fragile weathering.
Coal seam	3 coal	0.5~0.69 (Average 0.6)		It generally contains 0-3 layers, 0.01-0.02m thick, and the lithology is carbonaceous mudstone and mudstone.
Immediate floor	Fine grained sandstone	1.2~2.8		Grey fine grained sandstone, argillaceous cement.
Basic floor	siltstone	1.5~3.0		Gray siltstone with horizontal bedding.

Figure 1. Top and bottom rock column.

The coal layer position is stable, and the structure is simple. There is a layer of 0.3–0.8 m mudstone between the No. 3 coal seam and the immediate roof. The direct base composition is fine-grained sandstone. The working face is arranged along an inclined direction for mining, and the goaf roof is managed using comprehensive mechanized mining and full caving methods. The inclination of the coal seam is 0°~3°, and the average is 1°. The occurrence characteristics of the No. 3 coal seam are illustrated in Figure 2.

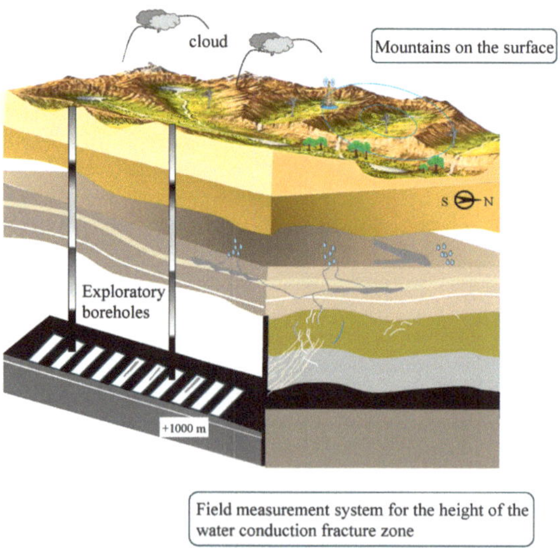

Figure 2. Occurrence location of No. 3 coal seam.

3. Construction of Data Set on the Height of the Water-Conduction Fracture Zone

3.1. Construction of the Index System of the Height of the Water-Conducting Fracture Zone

Figure 3 illustrates that after coal seam mining, the original stress balance is disrupted, leading to subsequent damage to the overlying rock. According to the failure characteristics, the overlying rock is divided into a caving zone, a fracture zone, and a bending subsidence zone from top to bottom, with the first two being referred to as water-conducting fracture zones. With the advancement of the working face, the water-conducting fracture zone undergoes a developmental process from occurrence, rise, reaching maximum height, falling back, and eventually stabilizing. According to the final shape of the overburden failure range formed after the advancement of the working face to a certain extent, the vertical distance between the highest point of the development of the water-conducting fracture zone and the upper mining boundary is considered as the maximum height of the water-conducting fracture zone, referred to as the guiding height [33].

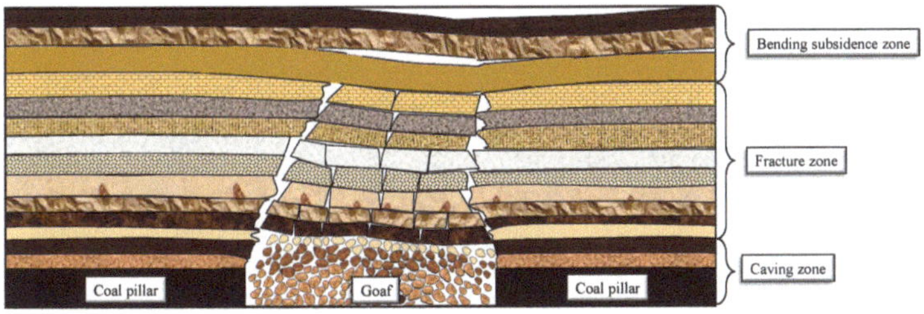

Figure 3. Feature map of three overburden zones.

The water-conducting fracture zone is constrained by multiple factors, but the empirical "Three Rules" take into account only a few influencing factors. Based on previous research results [21], this paper conducts a quantitative analysis of multiple factors.

(1) Mining thickness
Mining thickness affects the stress redistribution of roof strata and the failure range of overburden rock. It is the most critical factor for the failure of overburden rock and the development height of water-conducting fracture zones. In general, the greater the mining thickness, the higher is the height of the caving zone and water-conducting fracture zone, and the ratio is approximately linear [34].

(2) Width of the working face
The width of the working face, which refers to the inclined layout distance of the working face, is an influential factor in determining the size of the mining space affecting the water-conducting fracture zone and the area of the mining space. In the initial stage of coal seam mining, the height of the water-conducting fracture zone gradually increases with the increase in working face width [35].

(3) Mining Depth
The buried depth of a coal seam alters the initial stress of overlying rock, leading to the failure of the overlying rock under compression, tension, and shear stress. The greater the buried depth of a coal seam, the greater is the stress on the overlying rock, leading to a corresponding increase in the degree of damage to the overlying rock [34].

(4) Overburden Lithology
The development of mining fractures in overlying rock is closely related to the mechanical properties of the overlying rock, which determine the strength of the rock layer. Brittle rock layers are prone to fracturing, while plastic rock layers are more likely to bend and deform without developing fractures easily. The harder the rock layer, the greater is the height of overlying rock failure and the water-conducting fracture zone [34–37].

(5) Coal Seam Inclination
The change in the inclination angle of a coal seam mainly affects the failure pattern of overlying rock. The water-conducting fracture zone of a horizontal coal seam has a saddle shape with high ends and a low middle. In contrast, the caving rock of an inclined coal seam fills the lower goaf due to gravity, resulting in the water-conducting fissure zone taking on an upper and lower shape [34].

(6) Mining Methods
The influence of the mining method on the height of the water-conducting fracture zone mainly depends on the size of the mining space and the movement form of caving rock in the goaf [37].

In the study of water-conducting fracture zones under fully mechanized mining, both domestic and foreign scholars have identified mining thickness, mining depth, working face width, coal seam inclination angle, and overlying rock lithology as the primary factors influencing the height of water-conducting fracture zones.

3.2. Data Set Construction

By collecting a large number of field-measured data, we summarized a total of 35 examples of the height of water-conducting fracture zones in extremely thin coal seams, as shown in Table 1. According to the examples summarized in Table 1, the maximum mining depth (H) is 145 m, and the minimum mining depth (H) is 50 m. The maximum value of coal seam inclination, A, is 3°, and the minimum value is 0°. The maximum thickness (M) of the coal seam is 0.85 m, while the minimum thickness is 0.4 m. The maximum value of the working face width (L) is 170 m, while the minimum value is 100 m. The hardness of a coal seam is classified according to the Platts coefficient. The hardness of the No. 3 coal seam in the Niuwu Mining area is soft coal, with a value of 0.4. Previous studies have categorized the rock structure based on the lithology characteristics of the overlying rock into hard–hard, soft–hard, hard–soft, and soft–soft, with values of 1, 0.75, 0.5, and 0.25, respectively. The above data are normalized according to the following formula, and the input and output data are converted into interval (−1,1) dimensional values. The normalization processing formula x_n [38] is:

$$x_n = \frac{2(X_n - X_{\min})}{X_{\max} - X_{\min}} - 1, \tag{1}$$

where: x_n is the nth normalized data of various parameters; X_n is the nth data of each parameter. $X_{\max}$ and $X_{\min}$ represent the maximum and minimum values of each parameter.

Table 1. Measured values and influencing factors of the mining overburden height.

Number	Face Drilling	Mining Thickness/m	Mining Depth/m	Dip Angle/°	Width/m	Rock Structure	Height of Water-Conduction Fracture Zone/m
1	GJW1303	0.57	67	2	119	hard–soft	9.7
2	HH3201	0.64	50	1	110	hard–soft	7.3
3	AZ1201	0.55	30	3	100	hard–soft	10.2
4	HC1107	0.84	50	3	142	hard–soft	9.5
5	HC2	0.85	76	2	142	hard–soft	10.6
6	AZ1	0.61	70	1	100	soft–hard	7.3
7	AZ2	0.55	82	3	100	soft–hard	6.9
8	N1	0.60	60	1	120	hard–soft	8
9	N3	0.70	59	3	128	soft–soft	8.7
10	N5	0.40	79	2	128	soft–soft	6.1
11	KN1	0.70	145	0	130	hard–soft	8.5
12	M1	0.58	123	3	119	hard–soft	6.6

Table 1. *Cont.*

Number	Face Drilling	Mining Thickness/m	Mining Depth/m	Dip Angle/°	Width/m	Rock Structure	Height of Water-Conduction Fracture Zone/m
13	M7	0.54	106	1	120	soft–hard	6.7
14	M9	0.57	95	2	125	soft–hard	7.2
15	X1	0.55	58	2	150	hard–soft	6.9
16	N2	0.62	90	2	140	hard–soft	7.7
17	C1	0.63	67	1	170	soft–hard	8.2
18	C2	0.59	72	3	170	soft–hard	7.4
19	N2	0.51	65	3	120	hard–soft	6.5
20	C3	0.66	95	3	170	soft–hard	8.6
21	N6	0.50	75	2	140	hard–soft	9.2
22	F5	0.67	100	1	135	soft–soft	8.4
23	G1	0.53	103	2	121	soft –soft	6.5
24	G2	0.69	85	1	123	soft–hard	8.8
25	G3	0.49	90	3	125	hard–soft	8.4
26	G4	0.61	65	2	120	hard–hard	7.9
27	G5	0.66	71	2	140	hard–hard	8.2
28	A1	0.58	78	1	145	soft–hard	7.6
29	A2	0.72	80	1	100	soft–hard	9.1
30	A3	0.64	87	1	142	soft–soft	8.8
31	H1	0.55	90	1	125	hard–soft	8.4
32	H2	0.67	108	2	155	soft–hard	9.1
33	Q5	0.59	82	3	145	soft–soft	7.9
34	X5	0.49	75	2	145	hard–hard	8.7
35	H3	0.69	120	2	160	hard–soft	10

Normalization is a common data preprocessing method whose main purpose is to scale the data to the same range and eliminate dimensional differences between features. Dimensional difference refers to the difference in the value range and units of different features, which affects the performance and stability of the model. Through normalization processing, the value range of different features can be scaled to the same interval, so as to eliminate the impact of dimensional differences and improve the performance and stability of the model.

Normalization is widely used in machine learning to improve the performance and stability of models. In feature engineering, normalization can be used to scale the value range of different features to the same interval to improve the performance and stability of the model. In image processing, normalization can be used to scale the pixel value of the image to the range of (0,1), which is convenient for subsequent processing and analysis. In natural language processing, normalization can be used to convert text data into numerical vectors for easy processing and analysis by machine learning algorithms.

The data in Table 1 were normalized, and the processing results are presented in Table 2.

Table 2. Sample data processing results.

Number	Face Drilling	Mining Thickness/m	Mining Depth/m	Dip Angle/°	Width/m	Rock Structure	Height of Water-Conduction Fracture Zone/m
1	GJW1303	−0.244	−0.642	0.333	−0.457	−0.333	0.600
2	HH3201	0.067	−1.000	−0.333	−0.714	−0.333	−0.467
3	AZ1201	−0.333	0.684	1.000	−1.000	−0.333	0.822
4	HC1107	0.956	−1.000	1.000	0.200	−0.333	0.511

Table 2. *Cont.*

Number	Face Drilling	Mining Thickness/m	Mining Depth/m	Dip Angle/°	Width/m	Rock Structure	Height of Water-Conduction Fracture Zone/m
5	HC2	1.000	−0.453	0.333	0.200	−0.333	1.000
6	AZ1	−0.067	−0.579	−0.333	−1.000	0.333	−0.467
7	AZ2	−0.333	−0.326	1.000	−1.000	0.333	−0.644
8	N1	−0.111	−0.789	−0.333	−0.429	−0.333	−0.156
9	N3	0.333	−0.811	1.000	−0.200	−1.000	0.156
10	N5	−1.000	−0.389	0.333	−0.200	−1.000	−1.000
11	KN1	0.333	1.000	−1.000	−0.143	−0.333	0.067
12	M1	−0.200	0.537	1.000	−0.457	−0.333	−0.778
13	M7	−0.378	0.179	−0.333	−0.429	0.333	−0.733
14	M9	−0.244	−0.053	0.333	−0.286	0.333	−0.511
15	X1	−0.333	−0.832	0.333	0.429	−0.333	−0.644
16	N2	−0.022	−0.158	0.333	0.143	−0.333	−0.289
17	C1	0.022	−0.642	−0.333	1.000	0.333	−0.067
18	C2	−0.156	−0.537	1.000	1.000	0.333	−0.422
19	N2	−0.511	−0.684	1.000	−0.429	−0.333	−0.822
20	C3	0.156	−0.053	1.000	1.000	0.333	0.111
21	N6	−0.556	−0.474	0.333	0.143	−0.333	0.378
22	F5	0.200	0.053	−0.333	0.000	−1.000	0.022
23	G1	−0.422	0.116	0.333	−0.400	−1.000	−0.822
24	G2	0.289	−0.263	−0.333	−0.343	0.333	0.200
25	G3	−0.600	−0.158	1.000	−0.286	−0.333	0.022
26	G4	−0.067	−0.684	0.333	−0.429	1.000	−0.200
27	G5	0.156	−0.558	0.333	0.143	1.000	−0.067
28	A1	−0.200	−0.411	−0.333	0.286	0.333	−0.333
29	A2	0.422	−0.368	−0.333	−1.000	0.333	0.333
30	A3	0.067	−0.221	−0.333	0.200	−1.000	0.244
31	H1	−0.333	−0.158	−0.333	−0.286	−0.333	0.022
32	H2	0.200	0.221	0.333	0.571	0.333	0.333
33	Q5	−0.156	−0.326	1.000	0.286	−1.000	−0.200
34	X5	−0.600	−0.474	0.333	0.286	1.000	0.156
35	H3	0.289	0.474	0.333	0.714	−0.333	0.733

4. Multiple Regression Prediction Model

Based on previous research results [39–41], it has been found that the development of water-conducting fracture zones in mining overlying rock is not solely influenced by a single factor. The development of conductivity height is restricted by multiple factors, and each factor exhibits a linear relationship with the water-conducting fracture zone. The empirical formula of the "Three Rules" considers only a few factors, leading to limited adaptability and ineffective application of the prediction model. Based on correlation analysis, we selected mining thickness, mining depth, dip angle, working face width, and overburden lithology as multiple regression indicators, as described in Table 3. The model was established, and the prediction results were analyzed.

Table 3. Analysis of the correlation between each factor and the water-conducting fracture zone.

		Rock Structure	Mining Thickness/m	Mining Depth/m	Dip Angle/°	Width/m
Height of water-conduction fracture zone/m	Pearson correlation	0.395 *	0.607 **	0.419 *	0.376 *	0.391 *
	Significance (double tail)	0.031	0.000	0.021	0.041	0.033
	N	30	30	30	30	30

Notes: * The correlation is significant when the confidence (double test) is 0.05; ** The correlation is significant when the confidence (double measure) is 0.01.

Table 3 illustrates that at a significance level of 0.05, the height of the water-conducting fracture zone is significantly positively correlated with the rock structure, mining depth, dip angle, and width. The correlation coefficients are 0.395, 0.419, 0.376, and 0.391, respectively, with corresponding significance levels of 0.031, 0.021, 0.041, and 0.033, all of which are less than 0.05. When the significance level is 0.01, there is a significant positive correlation between the height of the water-conducting fracture zone and the thickness of the coal seam. The correlation coefficient is 0.607, and the significance is 0, which is less than 0.01.

According to the correlation analysis of the above five factors, mining thickness has the greatest influence on the water-conducting fracture zone, followed by mining depth, overlying rock lithology, working face width, and finally the dip angle. It can be seen that the influence of the five factors on the water-conducting fracture zone is also strong and weak.

4.1. Modeling Analysis of Multiple Linear Regression Model

Multiple linear regression involves two or more influential factors as independent variables to explain changes in the dependent variable. When there is a positive correlation between the five factors and the height of the water-conducting fracture zone, the correlation coefficient is positive, and the regression coefficient is also positive, indicating that an increase in the independent variable will lead to an increase in the dependent variable. When there is a negative correlation between the independent variable and the dependent variable, the correlation coefficient is negative, and the regression coefficient is also negative, indicating that an increase in the independent variable will lead to a decrease in the dependent variable. Based on the results obtained in Table 3, the correlation of various factors was analyzed. The five factors were all positively correlated with the height of the water-conducting fracture zone; therefore, the five factors, namely mining thickness, mining depth, dip angle, working face width, and overlying rock lithology, were adopted for modeling.

The five factors were taken as horizontal coordinates, while the development height of the water-conducting fracture zone was taken as the vertical coordinate to create scatter plots (refer to Figure 4).

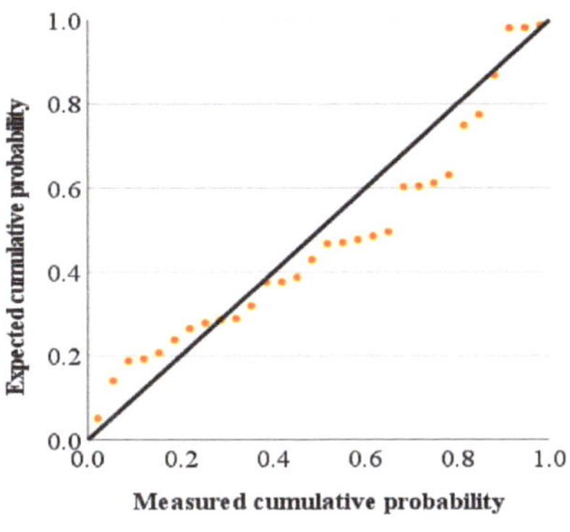

Figure 4. Normal P-P plot of standardized residuals.

This method is used to determine whether the five factors have a linear relationship with the height of the water-conducting fracture zone. The collected 30 sample data are in good agreement with the trend line in Figure 4, indicating a linear positive correlation between the various factors and the development height of the water-conducting fracture zone.

SPSS statistical software was used to conduct linear regression analysis on mining thickness, mining depth, dip angle, working face width, overburden lithology, and water-conducting fracture zone. The results of the regression analysis are as follows.

Table 4 presents the F-test results of the regression equation, and the significance value of $p = 0.000$ (<0.01) indicates that the linear regression equation is statistically significant.

Table 4 also presents the T-test results of the regression coefficient, indicating a proportional relationship between various factors and the development height of the water-conducting fracture zone.

Table 4. Analysis of variance of sample data.

Model	Sum of Squares	Degrees of Freedom	Mean Square	F	Significance
Regression	36.818	5	7.364	104.171	0.000
Residual error	1.697	24	0.071		
Total	38.515	29			

Notes: 1 Dependent variable: height of water-conducting fracture zone/m; 2 Predictive variables: (constant), working face width/m, dip angle/°, mining depth/m, mining thickness/m, rock structure.

Multiple linear regression is a statistical method used to examine the relationship between a dependent variable (Y) and multiple independent variables (X_1, $X_2 \ldots X_n$) through a linear regression equation. It helps to understand the relationship between the variables. The expression is given by Equation (2):

$$Y = a_0 + a_1x_1 + a_2x_2 + \cdots + a_kx_k + b_0, \quad (2)$$

In Equation (2), where b_0 is the constant term, $a_1, a_2 \ldots\ldots a_k$ are the partial regression coefficients for Y, and $x_1, x_2 \ldots x_k$ are the arguments.

According to the results presented in Table 5 and the significance of each factor, the five multifactorial variables of mining thickness (M), mining depth (H), dip angle (A), working face width (L), rock structure (J), and the height of the water-conducting fracture zone were utilized to analyze the 30 data points collected in Table 1. SPSS was employed for conducting multivariate linear regression analysis on the aforementioned data. The fitting results are shown in Equation (3):

$$Y_1 = 2.673M + 0.009H + 0.491A + 0.036L - 1.132J; \quad (3)$$

where M represents the thickness in meters, H represents the mining depth in meters, A represents the angle in degrees, L represents the width in meters, and J represents the rock structure. According to relevant statistical knowledge, the closer the goodness-of-fit measure R is to 1, the higher is the degree of fit and the better is the model's effectiveness. The model R = 0.629 indicates that the prediction model for water-conducting fracture zones has a general effect, and the prediction accuracy aligns well with the measured results. This alignment is primarily attributed to the limited amount of data available.

Table 5. Undetermined coefficient obtained by sample fitting.

Model		Unnormalized Coefficient		Standardization Coefficient	t	Significance
		B	Standard Error	Beta		
1	(constant)	0.938	1.453		0.645	0.525
	Mining thickness/m	2.673	1.094	0.207	2.443	0.022
	Mining depth/m	0.009	0.004	0.149	2.153	0.042
	Dip Angle/°	0.491	0.129	0.303	3.818	0.001
	Width/m	0.036	0.012	0.237	3.069	0.005
	Rock structure	−1.132	0.369	−0.218	−3.067	0.005

Note: 1 Dependent variable: height of water-conducting fracture zone/m.

4.2. Modeling and Analysis of Multiple Regression Nonlinear Model

Multivariate nonlinear regression prediction models aim to fit a nonlinear function to sample data and then use the fitting function to predict the response value. Commonly used function forms include power functions, exponential functions, and logarithmic functions.

Among them, the power function exhibits strong generalization ability, and the prediction model selects the power function as the form of its prediction model. The basic form is as follows:

$$Y = b_0 + b_1 x_1^{c_1} + b_2 x_2^{c_2} + \cdots\cdots b_3 x_1 x_2 + \cdots\cdots. \tag{4}$$

Since the model is based on the formation conditions in Fuxian County, northern Shaanxi Province, it is analyzed according to the occurrence of coal seams in the study area. Based on the method of multivariate nonlinear regression analysis, SPSS software was used to establish the regression model, and the variables were modeled sequentially. The mathematical model constructed is as follows:

$$Y_2 = a_0 M + a_1 M^2 + b_1 H + b_2 H^2 + b_3 H^3 + c_1 J + c_2 J^2 + c_3 J^3 + b_0, \tag{5}$$

$$Y_3 = a_0 M + a_1 M^2 + b_1 H + b_2 H^2 + b_3 H^3 + c_1 J + c_2 J^2 + c_3 J^3 + b_0, \tag{6}$$

$$Y_4 = a_0 M + a_1 M^2 + b_1 H + b_2 H^2 + b_3 H^3 + c_1 J + c_2 J^2 + c_3 J^3 + b_0, \tag{7}$$

where M is the mining thickness, H is the mining depth, and J is the rock structure.

The model was calculated by SPSS software, and 27 parameters of Equations (5)–(7) above needed to be fitted. $a_0, a_1, b_1, b_2, b_3, c_1, c_2, c_3$, and b_0 in the Equations citation (5)–(7) above are the factor coefficients in the three formulas, respectively, and the parameter fitting results are shown in Table 6 below.

Table 6. Fitting parameters of the prediction model of the water-conducting fracture zone.

	Reference Parameter								
1°	a_0 10.108	a_1 0	b_0 −17.385	b_1 0.508	b_2 −0.005	b_3 1.305×10^{-5}	c_1 12.47	c_2 −13.195	c_3 0
2°	a_0 5.79	a_1 0	b_0 −17.659	b_1 0.381	b_2 −0.002	b_3 0	c_1 44.046	c_2 −73.461	c_3 36.749
3°	a_0 −30.162	a_1 29.343	b_0 12.781	b_1 0.055	b_2 0	b_3 0	c_1 −1.664	c_2 0	c_3 0

The obtained multivariate nonlinear regression fitting formula is shown in the following Equations (8)–(10):

$$Y_2 = 10.108M - 17.385 + 0.508H - 0.005H^2 + 1.305E - 5H^3 + 12.47J - 13.195J^2 \tag{8}$$

$$Y_3 = 5.79M + 0.381H - 0.002H^2 + 44.046J - 73.461J^2 + 36.749J^3 - 17.659 \tag{9}$$

$$Y_4 = 12.781 - 30.162M + 29.343M^2 + 0.055H - 1.664J \tag{10}$$

R^2 was used to evaluate the model's effectiveness after fitting. After calculations, the corrected R^2 values of 0.923, 0.759, and 0.435 under three different inclination angles are all high, indicating a strong fitting effect.

4.3. *Verification of Multiple Regression Prediction Model*

According to the multiple linear regression prediction model and multiple nonlinear regression prediction model established above, the predicted values of Y_1, Y_2, Y_3, and Y_4 water-conducting fracture zones (Y_1, Y_2, Y_3, and Y_4) in overlying rocks can be obtained under specific mining conditions.

The predicted value obtained based on the multiple linear regression prediction model was compared with the empirical formula of the Three Rules. Figure 5 illustrates that the trend change of the multiple linear regression prediction model is consistent with the measured value of the water-conducting fracture zone. The maximum absolute error is 4.5 m, and the maximum relative error is 58%. The multiple nonlinear regression prediction model remains consistent with the measured value of the water-conducting fracture zone. The maximum absolute error is 2.1 m, and the maximum relative error is 23%, and the overall trend is the same. The error of the predicted value obtained by the three-part procedure is relatively large, and the practicability for extremely thin coal seams is low. Although the prediction accuracy of the multiple linear regression model is lower than that of the multi-part nonlinear regression model, its accuracy is still higher than that of the three-part procedure. It shows that both the multiple linear regression model and the multiple nonlinear regression model can effectively predict the height of the water-conducting fracture zone in very thin coal seams. The above prediction model can obtain the corresponding prediction value of the development height of the water-conducting fracture zone by inputting the formation parameters of very thin coal seams, and more accurate prediction value can be obtained by the multi-factor index, which indicates that the prediction method has high universality.

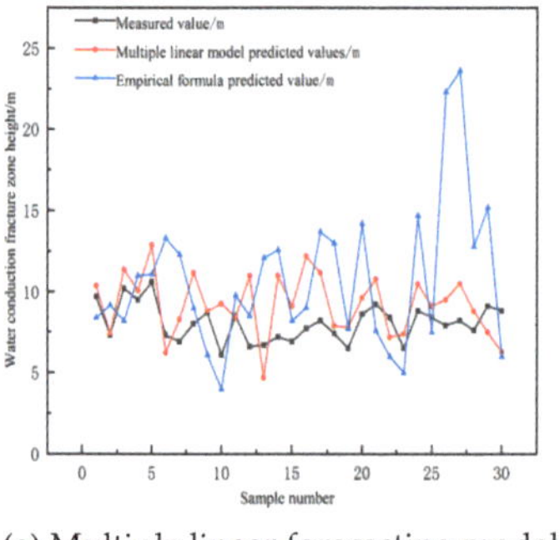

(**a**) Multiple linear forecasting models

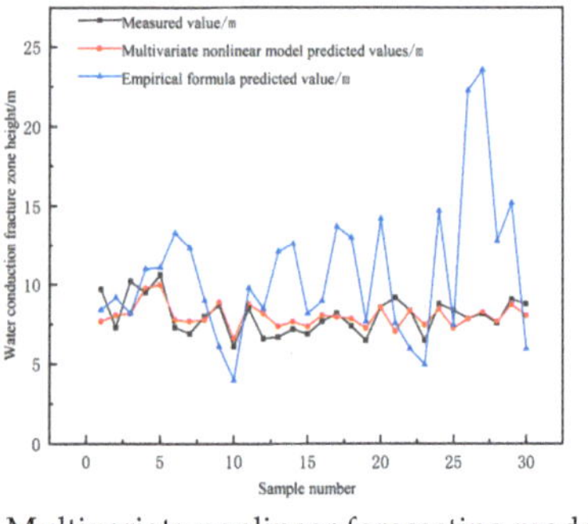

(**b**) Multivariate nonlinear forecasting models

Figure 5. Comparison of the predicted and measured values of the development height of the water-conducting fracture zone.

5. BP Neural Network Model

5.1. Construction of BP Neural Network Model

The BP neural network is one of the widely used neural network models at present. The neural network model creates a multi-dimensional input and output function based on the training sample and utilizes this function to make predictions. The training process of the network involves adjusting the parameters of the function to enhance prediction accuracy. Due to its strong nonlinear mapping capability, self-learning, and generalization, artificial intelligence is extensively utilized in data processing, machine control, and pattern recognition.

According to the principle of sample selection for BP neural network construction, it is necessary to quantify the factors affecting the development height of water-conducting fracture zones. Based on the measured data on the development height of the water-conducting fracture zone of an extremely thin coal seam in Fuxian County, northern Shaanxi Province, 35 field measured data were selected as the learning training samples and test samples of the BP neural network (see Table 1), of which 30 were trained samples. Five test samples were used to evaluate the generalization ability of the final prediction model. When building the BP neural network model, we used Python language to write programs for simulation training and build a Pytorch framework to realize the BP neural network prediction model, and we used Pytorch to build a framework to further use Python language for quantitative data processing and intended to use mean squared error (MSE) as a loss function. The rectified linear unit (ReLu) was processed as an incentive function. The general steps of the BP neural network algorithm are shown in Figure 6.

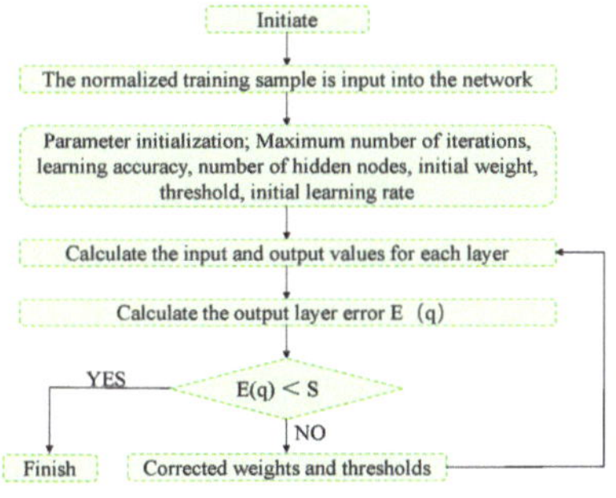

Figure 6. BP neural network algorithm steps.

This paper aims to utilize the enhanced three-layer BP model with one hidden layer. The basic structure is as follows: the input layer consists of five nodes, namely, mining thickness (*M*), mining depth (*H*), coal seam inclination (*A*), working face width (*L*), and rock structure (*J*). The output layer comprises one node, indicating the height (H_f) of the water-conducting fracture zone. For the hidden layer, the number of nodes is selected as 128 based on multiple training and experience calculations. The structure of the BP neural network model is shown in Figure 7. Through the continuous use of "signal forward calculation" and "error back propagation" reciprocating until the global error approaches the minimum value, the purpose of the prediction is finally achieved.

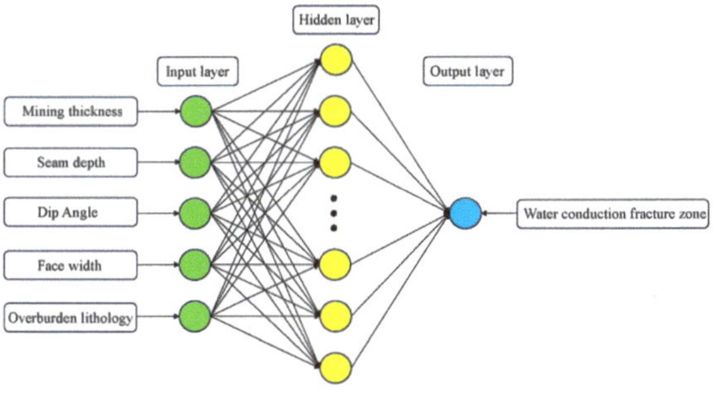

Figure 7. Structure of the neural network model for the calculation of the height of the water-conduction fracture zone.

5.2. Network Learning and Training

The parameters of the model were conservatively selected to prioritize the stability of the model over training speed, so the three- and four-layer BP neural network were used as the training models. The loss function selected was the MSE function, and the activation function chosen was the ReLu function.

MSE is a commonly used loss function in regression tasks. It measures the average squared error between the predicted value and the actual value of the model. The specific formula is as follows:

$$MSE = \frac{1}{n}\sum (x_i - y_i)^2, \tag{11}$$

where n is the number of samples, x_i is the true value of the ith sample, and y_i is the predicted value of the model for the ith sample.

The ReLu function is a type of activation function commonly utilized in artificial neural networks. In a mathematical sense, it refers to the slope function, namely:

$$f(x) = \max(0, x), \tag{12}$$

The corresponding function image is shown in Figure 8.

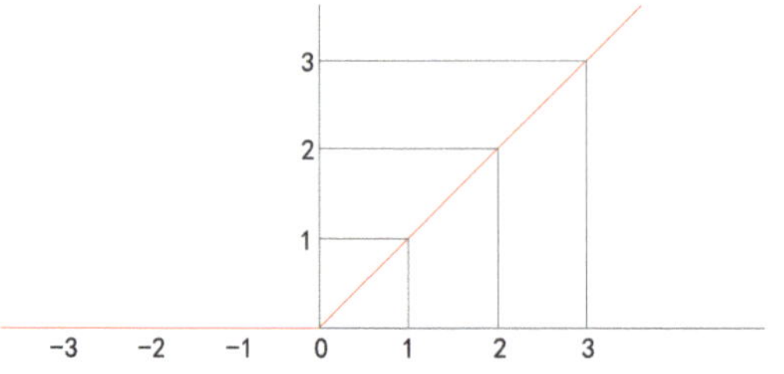

Figure 8. Excitation function (ReLu function) image.

The initial learning rate was set to 0.0001, and MSE, R-squared determination coefficient, training duration, and model complexity were chosen as evaluation metrics. The results are shown in Table 7 below.

Table 7. Parameters for training the model.

Model Name	Input Layer	Hidden Layer	Output Layer
3-layer BP neural network	1	1	1
4-layer BP neural network	1	2 (128 neurons/layer)	1
5-layer BP neural network	1	3 (128 neurons/layer)	1

The core idea of support vector regression (SVR) is to find an optimal regression hyperplane in the high-level space by mapping the input space. The goal of SVR is to balance model complexity and error, and finally obtain a regression model with good generalization ability. Its mathematical expression is:

$$f(x) = \omega x + b \tag{13}$$

To ensure that the prediction error of all data points (*xi*, *yi*) is within ε, while minimizing ||ω||2, or expressed as $\min\frac{1}{2}\|\omega\|^2$, its constraint conditions are:

$$\begin{aligned} &\left|y_i - (\omega x_i + b)\right| \le \varepsilon \\ &\xi_i \ge 0, \xi_i^{※} \ge 0 \end{aligned} \tag{14}$$

The relaxation variables of type ξ_i and $\xi_i^{※}$ are used to process data points that cannot fall within ε.

By introducing the concept of kernel function $K(v1, v2)$, the complex inner product calculation of high dimensional space is transformed into the substitution calculation of a general function. Then, according to the principle of minimum classification interval, a regression model is established:

$$f(x) = \sum_i^j (\alpha_i^{※} - \alpha_i) K(x_i, x) + b, \tag{15}$$

$$b = y_j - \sum_i^j (\alpha_i - \alpha_i^{※})(x_i, x_j) + \varepsilon, \tag{16}$$

where ε is the precision coefficient. The training samples whose Lagrange coefficient $(\alpha^{※} - \alpha)$ is not 0 are the "support vectors". The regression model is then used for data fitting and prediction. The diagram is shown in Figure 9.

7. Conclusions

(1) The multiple regression linear model and the multiple regression nonlinear model of extremely thin coal seam show that the trend change between the measured value and the predicted value is consistent, and the error range is obviously reduced with the formula in the Three Gorges regulation, indicating that the regression prediction model is effective in predicting the height of the water-conducting fracture zone. It provides theoretical guidance for the field measurement of water-conducting fracture zones.

(2) Through comparative analysis, multiple linear regression, multi-distance nonlinear regression, and BP neural network can all improve the prediction accuracy of the height of the water-conducting fracture zone. Among several prediction methods, the BP neural network had the best effect.

(3) The BP neural network, multiple linear regression model, multi-distance nonlinear regression model, and traditional three-order empirical formula were compared to analyze their relative error and absolute error. The BP neural network has good fault tolerance by using "error backpropagation", reducing the correlation between factors and improving the degree of model prediction. Moreover, the prediction accuracy of the BP neural network is as high as 93%, and the predicted result is basically consistent with the real value, which proves the accuracy of the model, and the application effect is good. This method can be used to predict the height of mine water-conduction fracture zones and guide mine safety production.

(4) Since there is limited research on mining extremely thin coal seams both domestically and internationally, the height of the water-conducting fracture zone will be influenced by various factors during coal seam mining. Since our research focuses on a single mining area, the considered factors are relatively limited. In the future, it is advisable to explore additional influencing factors and conduct quantitative analysis using various modified optimal models. By doing so, the accuracy of the optimal model will be improved, providing a reference value for predicting the development height of water-conducting fracture zones under mining conditions of extremely thin coal seams.

Author Contributions: Conceptualization, H.W. and L.L.; methodology, H.W. and J.T.; data curation, H.W., D.C. and Y.Y.; writing—original draft preparation, J.T.; writing—review and editing, H.W., J.T., B.L. and D.C.; funding acquisition, Y.Y. All authors have read and agreed to the published version of the manuscript.

Funding: This research was funded by the National Natural Science Foundation of China, grant number 51974231.

Data Availability Statement: The authors confirm they have included a data availability statement in their main manuscript file. The datasets generated during and/or analyzed during the current study are available from the corresponding author upon reasonable request.

Acknowledgments: A special acknowledgment should be shown to the anonymous reviewers for their constructive and valuable comments. We thank them for taking time from their busy schedule to provide guidance.

Conflicts of Interest: The authors declare no conflicts of interest.

References

1. Gao, Y. "Four-belt" model of rock displacement and dynamic displacement inverse analysis. *J. China Coal Soc.* **1996**, 51–56.
2. Qian, M.; Xu, J.; Miao, X. Green mining technology of coal mine. *J. China Univ. Min. Technol.* **2003**, 5–10.
3. State Administration of Coal Industry. *Regulations for Coal Pillar Retention and Coal Mining in Buildings, Water Bodies, Railways and Main Shafts and Lanes*; Coal Industry Press: Beijing, China, 2000.
4. Shi, L.Q.; Xin, H.Q.; Zhai, P.H.; Li, S.C.; Liu, T.B.; Yan, Y.; Wei, W.X. Research on height calculation of water-conducting fracture zone under large mining depth. *J. China Univ. Min. Technol.* **2012**, *41*, 37–41.
5. Li, Z.; Xie, J.; Wang, F. Position and mining height are broken key stratum of abutment pressure influence law of the research and application. *J. Coal Technol.* **2015**, *34*, 34–36.
6. Dong, Y.; Huang, Y.; Du, J.; Zhao, F. Study on Overburden Stability and Development Height of Water Flowing Fractured Zone in Roadway Mining with Cemented Backfill. *J. Shock. Vib.* **2021**, *2021*, 1–12. [CrossRef]

7. Xu, J.; Zhu, W.; Wang, X. Based on the position of key stratum water fractured zone height is expected method. *J. Coal* **2012**, *5*, 762–769.
8. Zha, H.; Liu, W.; Liu, Q. Physical Simulation of the Water-Conducting Fracture Zone of Weak Roofs in Shallow Seam Mining Based on a The Self-Designed Hydromechanical Coupling Experiment System. *J. Geofluids* **2020**, *2020*, 1–14. [CrossRef]
9. Sun, D.; Li, X.; Zhu, Z.; Li, Y.; Cui, F. Height of the Fractured Zone in Coal Mining under Extra-Thick Coal Seam Geological Conditions. *Adv. Civ. Eng.* **2021**, *2021*, 9998545. [CrossRef]
10. Zvarivadza, T. Sustainability in the mining industry: An evaluation of the National Planning Commission's diagnostic overview. *Resour. Policy* **2018**, *56*, 70–77. [CrossRef]
11. Onifade, M.; Zvarivadza, T.; Adebisi, J.A. Advancing toward sustainability: The emergence of green mining technologies and practices. *Green Smart Min. Eng.* **2024**, *1*, 157–174. [CrossRef]
12. Huang, Q. Shallow buried coal seam mining strata water control research. *J. Coal* **2017**, *42*, 50–55.
13. Lai, X.; Cui, F. Mining Pressure Distribution Lawand Disaster Prevention of lsolated lsland Working Face under the Condition of Hard "Umbrella Arch". *Rock Mech. Rock Eng.* **2024**, 1–19. [CrossRef]
14. Jie, Z.; Hou, Z. Physical simulation analysis of the development law of water-conducting fissure in shallow coal seam. *Mine Press. Roof Manag.* **2004**, 32–34+118.
15. Feng, J.; Wang, S.; Hou, E.; Ding, X.; Duan, H. Determining the Height of Water-Flowing Fractured Zone in Bedrock-Soil Layer in a Jurassic Coalfield in Northern Shaanxi, China. *J. Hindawi Ltd.* **2021**, *2021*, 9718802. [CrossRef]
16. Xu, H.; Lai, X.; Shan, P.; Yang, Y.; Zhang, S.; Yan, B.; Zhang, Y. Energy dissimilation characteristics and shock mechanism of coal-rock mass induced in steeply-inclined mining: Comparison based on physical simulation and numerical calculation. *Acta Geotech.* **2023**, *18*, 843–864. [CrossRef]
17. Yang, G.; Chen, C.; Gao, S.; Feng, B. Research on the development height of water-conducting fracture zone based on AHP and fuzzy clustering analysis. *J. Min. Saf. Eng.* **2015**, *32*, 206–212.
18. An, T.; Wang, L.; Pu, H.; Yao, B. Shallow buried coal seam water transmitting fissure zone development regularities of numerical simulation study. *J. Min. Res. Dev.* **2010**, *30*, 33–36+49.
19. Liu, Z.; Yang, B. Determination of the development height of water-conducting fracture zone by numerical simulation. *Min. Saf. Environ. Prot.* **2006**, 16–19+89.
20. Li, X. Numerical simulation of overburden failure height in mining area based on FLAC~(3D). *Coal Technol.* **2012**, *31*, 83–85.
21. Huang, W.; Gao, Y.; Wang, B.; Liu, J. Strata combination structure evolvement lead water fracture zones and development height analysis. *J. Min. Saf. Eng.* **2017**, *34*, 330–335.
22. Gu, H.; Lai, X.; Tao, M.; Momeni, A.; Zhang, Q. Dynamic mechanical mechanism and optimization approach of roadway surrounding coal water infusion for dynamic disaster prevention. *Measurement* **2023**, *223*, 113639. [CrossRef]
23. Lai, X.; Xu, H.; Shan, P.; Hu, Q.; Ding, W.; Yang, S. Research on mechanism of rock burst induced by mined coal-rock linkage of sharply inclined coal seams. *Int. J. Miner. Metall. Mater.* **2024**, *31*, 929–942. [CrossRef]
24. Dai, L.; Pan, Y.; Xiao, Y.; Wang, A.; Wang, W.; Wei, C.; Fan, D. Parameter Design Method for Destressing Boreholes to Mitigate Roadway Coal Bursts: Theory and Verification. *Rock Mech. Rock Eng.* **2024**, 1–18. [CrossRef]
25. Guo, X.; Liu, Z.; Gu, Z. His long mining area coal bed mining water flowing fractured zone height detection and calculation. *J. Control. Eng. Min. Strat.* **2023**, *5*, 91–100.
26. Lin, C.; Ren, D.; Wang, Q.; Li, C.; Wang, J.; Hang, H.; Liu, Q. The second panel of fort gaos thought field water fractured zone height measured research. *Coal Technol.* **2024**, *43*, 209–212.
27. Zhu, W.; Teng, Y.; Tang, Z. Lu'An mining height fully mechanized fissure zone development law of actual research. *J. Coal Sci. Technol.* **2017**, *45*, 167–171.
28. Li, Z.; Xu, Y.; Li, L.; Zhai, C. Of fracture zone height based on BP neural network prediction. *J. Min. Saf. Eng.* **2015**, *32*, 905–910.
29. Chai, H.L.; Zhang, J.; Yan, C. Based on GA-SVR mining-induced overburden rock lead water crack bring height prediction. *J. Min. Saf. Eng.* **2018**, *35*, 359–365.
30. Lou, G.; Tan, Y. Prediction of water conduction fracture zone height based on PSO-BP neural network. *Coal Geol. Explor.* **2021**, *49*, 198–204.
31. Shi, L.; Wu, H.; Li, Y.; Lv, W. A PCA-GA-Elman optimization model for predicting the development height of water-conducting fracture zone. *J. Henan Univ. Sci. Technol. (Nat. Sci. Ed.)* **2021**, *40*, 10–18.
32. Wu, Q.; Shen, J.; Liu, W.; Wang, Y. A RBFNN-based method for the prediction of the developed height of a water-conductivefractured zone for fully mechanized mining with sublevel caving. *Arab. J. Geosci.* **2017**, *10*, 1–9. [CrossRef]
33. Gao, B.; Liu, Y.; Pan, J.; Yuan, T. Mining under waterbodies feed water detection and analysis of the fractured zone height. *J. Rock Mech. Eng.* **2014**, *33*, 3384–3390.
34. Sheng, F.; Duan, Y. Research on the height of water-conducting fissure zone under full-mechanized caving in the giant thick sandstone aquifer in Binchang Mining area. *Coal Eng.* **2022**, *54*, 84–89.
35. Xu, S.; Zhang, Y.; Sun, H.; Hu, X. Research on prediction model of water conduction fracture zone height based on RBF core ε-SVR. *J. Saf. Environ.* **2021**, *21*, 2022–2029.
36. Li, B.; Wu, H.; Li, T. Research on weighted multivariate nonlinear regression prediction method for hydraulic fracture zone height in fully mechanized mining. *J. Min. Saf. Eng.* **2019**, *39*, 536–545.

37. Gui, Y. Research on the Height and Prediction Method of Hydraulic Fracture Zone in Fully Mechanized Caving Mining. Ph.D. Thesis, Shandong University of Science and Technology, Qingdao, China, 2004.
38. Hu, X.; Li, W.; Cao, D.; Liu, M. Fully mechanized water flowing fracture with multiple factors influence index research and highly expected. *J. Coal* **2012**, *5*, 613–620.
39. Xu, W.; Yin, S.; Xu, B.; Cao, M.; Zhang, R.; Tang, Z.; Huang, W.; Li, W. Fully mechanized working strata under the condition of water flowing fractured zone height prediction model optimization. *J. Coal Sci. Technol.* **2023**, *51*, 284–297.
40. Zhang, S.; Feng, G.; Zhang, D.; Wei, W. Yulin-shenmu coal mine areas water fractured zone height multi-factor influence law and water coal partition. *J. Taiyuan Univ. Technol.* **2023**, *54*, 301–312.
41. Zhai, Z.; Zhang, S.; Meng, X.; Wu, G.; Xu, F.; Xu, S. Research on highly comprehensive analysis technology for development of water-conducting fracture zone in overlying rock of coal seam. *Coal Eng.* **2022**, *54*, 112–116.

water

MDPI

Article

The Hydrochemical Characteristics and Formation Mechanism of Highly Mineralized Coal Mine Water in Semi-Arid Regions in Northwest China

Jian Yang [1], Wei Zhao [2,*], Xiangyang Liang [3] and Feng Xu [3]

[1] College of Geology and Environment, Xi'an University of Science and Technology, Xi'an 710054, China; yangjian0106@126.com
[2] School of Water and Environment, Chang'an University, Xi'an 710054, China
[3] Xi'an Research Institute of China Coal Technology & Engineering, (Group), Corp, Xi'an 710054, China
* Correspondence: zhaowei0471@126.com

Abstract: The over-exploitation of groundwater and the deterioration of its quality have heightened the importance of non-traditional water resources, such as mine water. The study of the water's chemical characteristics and the formation mechanism of high-salinity mine water in semi-arid regions holds significant importance for zero discharge and the resource utilization of mine water in Northwest China. In this study, a total of 38 groundwater and mine water samples were collected to examine the hydrogeochemical characteristics of high-salinity mine water using Piper diagrams and Gibbs diagrams, as well as isotope analyses and ion ratio coefficients. Additionally, the corresponding mine water treatment recommendations were put forward. The results show that the TDS content of groundwater increases with hydrographic depth. The average TDS concentration of Quaternary, Luohe, and Anding groundwater is 336.87, 308.67, and 556.29 mg/L, respectively. However, the TDS concentration of Zhiluo groundwater and mine water is 2768.57 and 3826.40 mg/L, respectively, which belong to high-salinity water. The Quaternary, Luohe, and Anding groundwater hydrochemical type is predominantly HCO_3-Ca type, and the Zhiluo groundwater and mine water hydrochemical type is predominantly the SO_4-Na type. Furthermore, there is minimal difference observed in δD and $\delta^{18}O$ values among these waters. It can be inferred that the Zhiluo Formation in groundwater serves as the primary source of mine water supply, primarily influenced by the processes of concentration caused by evaporation. The high salinity of mine water is closely related to the high salinity of Zhiluo groundwater. The high salinity of groundwater has evolved gradually under the control of the concentration caused by evaporation and rock-weathering processes. The dissolution of salt rock, gypsum, along with other minerals, serves as the material basis for high-salinity groundwater formation. In addition, the evolution of major ions is also affected by cation exchange. The TDS concentration of mine water ranges from 3435.4 mg/L to 4414.3 mg/L, and the combined treatment process of nanofiltration and reverse osmosis can be selected to remove the salt. After treatment, mine water can be used for productive, domestic, and ecological demands.

Keywords: high-salinity mine water; semi-arid regions; Northwest China; hydrogeochemical characteristics

Citation: Yang, J.; Zhao, W.; Liang, X.; Xu, F. The Hydrochemical Characteristics and Formation Mechanism of Highly Mineralized Coal Mine Water in Semi-Arid Regions in Northwest China. *Water* **2024**, *16*, 2244. https://doi.org/10.3390/w16162244

Academic Editor: Elias Dimitriou

Received: 4 July 2024
Revised: 29 July 2024
Accepted: 6 August 2024
Published: 8 August 2024

1. Introduction

The role of groundwater is indispensable in fulfilling essential human needs, serving as a critical water source for over 1.5 billion individuals worldwide [1,2]. However, the over-exploitation of groundwater and the deterioration of its quality have heightened the importance of non-traditional water sources, such as mine water [3]. As the world's largest coal producer, China is projected to produce 4.7 billion tons of raw coal in 2023, with demand expected to remain stable at approximately 4 billion tons per annum throughout the 14th Five-Year Plan period [4–6].

The shift in coal resource development to western China, driven by diminishing eastern reserves and increasing environmental constraints in central regions, has led to extensive groundwater seepage and the subsequent formation of mine water [7,8]. Each ton of raw coal extracted in China results in approximately 2.1 tons of mine water, with production projected to reach 800 million tons in 2023 [9]. This mine water often contains elevated levels of sulfate, fluoride, iron, manganese, and other contaminants, posing significant environmental and resource management challenges [10,11].

The ZYL Coal Mine in Yulin, Northwest China, serves as a case study to explore the fundamental formation mechanisms of high-salinity mine water. The primary source of mine water is derived from both surface water and groundwater, with high salinity primarily influenced by increased ion solubility [12–14]. Extensive research has been conducted on the distribution characteristics and change patterns of highly mineralized groundwater, with notable studies focusing on areas such as the Marcellus Basin, Ejina Basin, and Shiyang River Basin [15–17]. Researchers have also examined the distribution pattern of brackish water in northern China and the TDS concentration characteristics of high-salinity mine water in coal mining areas like Ningdong Coalfield and XinJulong Coalfield [18–21].

Analytical methodologies such as Piper's trilinear diagram, the ion ratio relationship diagram, and Gibb's semi-logarithmic diagram have been employed to study the hydrochemical characteristics and evolution processes of groundwater [22–25]. Additionally, advancements in the treatment of high-salinity mine water, including distillation, electrodialysis, and reverse osmosis, have been explored [26,27]. However, there has been limited focus on the chemical characteristics and mechanisms of high-salinity mine water.

This study aims to address this gap by investigating the ZYL Coal Mine in Yulin, Northwest China. The objectives are to (1) analyze the hydrogeochemical characteristics of high-salinity mine water, including the identification of major ions and trace elements, (2) elucidate the fundamental formation mechanisms of high-salinity mine water, encompassing geological, hydrological, and geochemical processes, and (3) propose effective treatment methods based on its hydrogeochemical characteristics and formation mechanisms, with a focus on enhancing water quality and minimizing environmental impact.

By providing a detailed investigation of the ZYL Coal Mine, this study contributes to a broader understanding of high-salinity mine water and offers insights applicable to similar contexts worldwide. The findings are expected to inform the development of more effective and sustainable management practices for high-salinity mine water in mining operations.

2. Method and Material

2.1. Study Area

The ZYL Coal Mine is situated in close proximity to the WS-Z Coal Mine and B-S1 Coal Mine to the south, as well as the K-G Coal Mine to the east (Figure 1). It is one of the major hundred-million-ton coal mines in the Northern Shaanxi Coal Mines, situated approximately 30 km northwest of Yulin City. The ZYL mining area covers a length of 14.7 km in the north–south direction and spans 12.5 to 23.3 km from east to west, encompassing an area of 265.63 km^2. The area falls within the administrative jurisdiction of Xiaojihan Township and Bulanghe Township in Yuyang District, Yulin City. The mine possesses a recoverable reserve of 3209.42 Mt and boasts a production capacity of 15.0 Mt/a. Recoverable coal seams include No. 2, No. 3, No. 4-2, No. 5, No. 6, No. 8, and No. 9, with particular emphasis on the main recoverable seams of No. 2 and No. 3 coal. The study area is situated on the southern periphery of the Mu Us Desert, characterized by extensive desert flats and partially stabilized dunes, with desert coverage exceeding 70%. The terrain exhibits higher elevation in the northeast and lower elevation in the southwest, ranging from 1200 to 1327.30 m. The annual precipitation ranges from 279 to 541 mm, with an average of 410 mm and average annual evaporation ranging from 1720 to 2085 mm. The location falls within the continental climate zone of the arid temperate plateau.

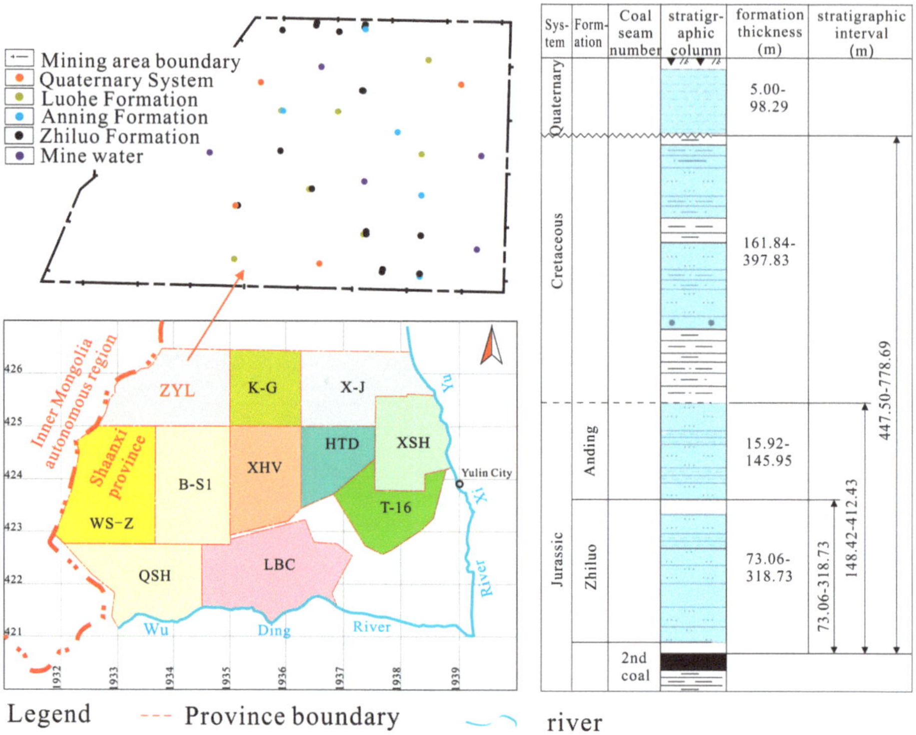

Figure 1. Location, water sampling, and hydrogeological histogram of study area.

The main strata in this region are arranged chronologically as follows: Quaternary, Neogene, Cretaceous, and Jurassic. The main aquifers are the pore aquifer of the Quaternary Salusu Formation, the pore (fissure) aquifer of the Cretaceous Luohe Formation, the pore (fissure) aquifer of the Anding Formation, and the fissure aquifer of the Zhiluo Formation, which are also the main direct or indirect water sources of the mine (Figure 1). The Quaternary aquifer is a phreatic layer located at a distance of 447.50~778.69 m from the No. 2 coal seam in the main mining area, exhibiting a permeability coefficient ranging from 0.479 to 8.42 $m \cdot d^{-1}$. The Cretaceous Luohe Formation aquifer is located at a distance of 148.42~412.43 m from the No. 2 coal seam in the main mining area. The aquifer has a thickness ranging from 80.91 to 397.83 m, with an average thickness of 273.42 m and a permeability coefficient ranging from 0.066 to 0.187 $m \cdot d^{-1}$. The Zhiluo Formation aquifer is located at a distance of 73.06~318.73 m from the main coal seam No. 2, exhibiting a thickness ranging from 38.60 to 256.19 m, with an average thickness of 106.19 m and a permeability coefficient ranging from 0.067 to 0.453 $m \cdot d^{-1}$.

2.2. Sample Collection and Testing

The hydrogeological and geochemical data of the ZYL Coal Mine are insufficient, and there is a lack of clarity regarding the water quality characteristics of each aquifer within the mining area. Thus, the hydrochemical characteristics of each aquifer in the minefield were investigated by collecting a total of 38 samples from the study area, including 4 samples of Quaternary groundwater, 7 samples of Luohe Formation water, 5 samples of Anding Formation water, 16 samples of Zhiluo Formation water, and 6 samples of mine water. The sampling containers were rinsed 3–5 times before collecting samples from each aquifer, and after sampling, the containers were promptly sealed without any air exposure. The sampling time and location were clearly marked on the sampling containers and subsequently dispatched to the laboratory for analysis. The tested parameters encompassed K^+, Na^+, Ca^{2+}, Mg^{2+}, SO_4^{2-}, Cl^-, HCO_3^-, total dissolved solids (TDSs), and δD and $\delta^{18}O$ isotopes. The ion concentrations and TDS levels were determined using standard analytical methods, including ion chromatography (IC) for major anions and inductively coupled plasma mass spectrometry (ICP-MS) for cations.

3. Results and Discussion

3.1. Hydrochemical Characteristics of Groundwater and Mine Water

The main ion concentrations of water samples from the ZYL Coal Mine are presented in Table 1. Analysis of various water samples from the study area revealed the following:

(1) The total dissolved solid (TDS) concentrations of Quaternary groundwater and the Anding Formation of Luohe Formation groundwater range from 216.99 to 911.767 mg/L, with an average of 393.10 mg/L.
(2) The TDS concentration of Zhiluo Formation groundwater and the No. 2 coal seam ranges from 1615.67 to 5124.0 mg/L, with an average value of 3057.07 mg/L.
(3) It can be inferred that the TDS content of groundwater increases with hydrographic depth.

Table 1. Main ion concentrations and TDSs of water samples from the ZYL Coal Mine ($mg{\cdot}L^{-1}$).

Water Sample Type	Statistical Item	$Na^+ + K^+$	Ca^{2+}	Mg^{2+}	SO_4^{2-}	HCO_3^-	Cl^-	TDS
Quaternary System	Maximum value	26.40	61.10	12.60	57.60	199.20	81.00	358.17
	Minimum value	1.14	3.05	0.30	20.37	122.00	5.30	292.00
	Average value	13.61	37.32	5.38	41.09	177.95	39.77	336.87
	Coefficient of variation	0.76	0.66	0.96	0.44	0.21	1.01	0.09
Luohe Formation	Maximum value	67.03	54.65	10.90	145.29	201.40	24.50	408.95
	Minimum value	6.81	19.95	5.20	21.00	133.98	1.15	216.99
	Average value	35.99	37.76	7.61	51.11	161.62	14.57	308.67
	Coefficient of variation	0.63	0.32	0.29	0.87	0.13	0.61	0.19
Anding Formation	Maximum value	211.78	139.52	12.20	513.15	210.50	78.43	911.76
	Minimum value	17.50	7.88	2.40	28.80	11.36	13.15	335.25
	Average value	100.50	58.20	6.72	230.96	121.91	38.00	556.29
	Coefficient of variation	0.71	0.87	0.58	1.01	0.63	0.66	0.50
Zhiluo Formation	Maximum value	1086.17	473.24	48.60	3442.00	146.40	76.10	5124.00
	Minimum value	371.80	80.20	4.90	946.20	55.68	0.00	1615.67
	Average value	566.99	279.60	18.82	1786.84	103.92	15.27	2768.57
	Coefficient of variation	0.31	0.53	0.74	0.40	0.27	1.39	0.36
Mine water	Maximum value	819.40	408.80	71.70	2896.20	142.10	46.10	4414.30
	Minimum value	476.50	124.30	10.90	2555.00	36.60	19.80	3435.40
	Average value	567.95	287.55	28.32	2711.82	111.40	35.62	3826.40
	Coefficient of variation	0.25	0.40	0.88	0.06	0.36	0.24	0.10

The dominant cations of water samples within the study area were primarily $Na^+ + K^+$, exhibiting a concentration range of 1.14 to 1086.17 mg/L with an average concentration of 349.70 $mg{\cdot}L^{-1}$, followed by Ca^{2+} and Mg^{2+}. The predominant anion is SO_4^{2-}, with a concentration range of 20.37 to 3442.00 mg/L and an average concentration of 1224.66 $mg{\cdot}L^{-1}$, followed by HCO_3^- and Cl^-.

From Table 1, the water samples in the study area are represented in a Piper diagram (Figure 2) [28]. The hydrochemical composition of the groundwater varies by formation: the Quaternary groundwater is of the HCO_3-Ca type, the Luohe Formation is of the HCO_3-Ca type and HCO_3-Na type, the Anding Group is of the HCO_3-SO_4-Na·Ca type, and the Zhiluo Formation and the mine water are of the SO_4-Na type. In addition, a Scholler diagram (Figure 3a) and a variable coefficient diagram (Figure 3b) are presented. The concentration of each ion in the fourth system, including Luohe and Anding groundwater samples, did not vary much. However, the SO_4^{2-}, $Na^+ + K^+$, Ca^{2+}, and Mg^{2+} concentration of Zhiluo groundwater and mine water is higher than that of Quaternary, Luohe, and Anding groundwater. In addition, the variable coefficients of the ions or TDS concentrations were calculated for quantitative dispersion of water quality (Figure 3b).

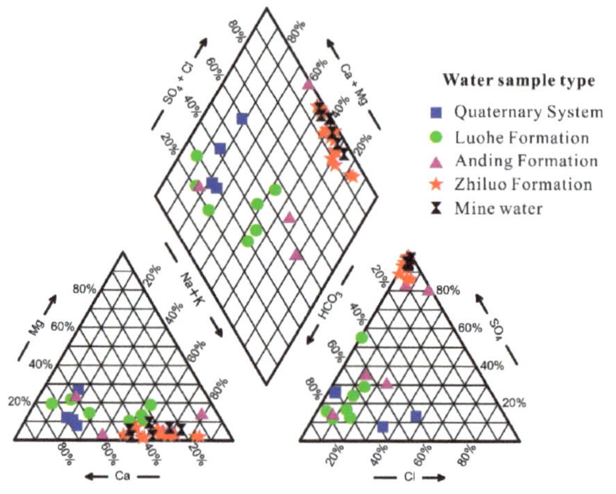

Figure 2. Piper diagram.

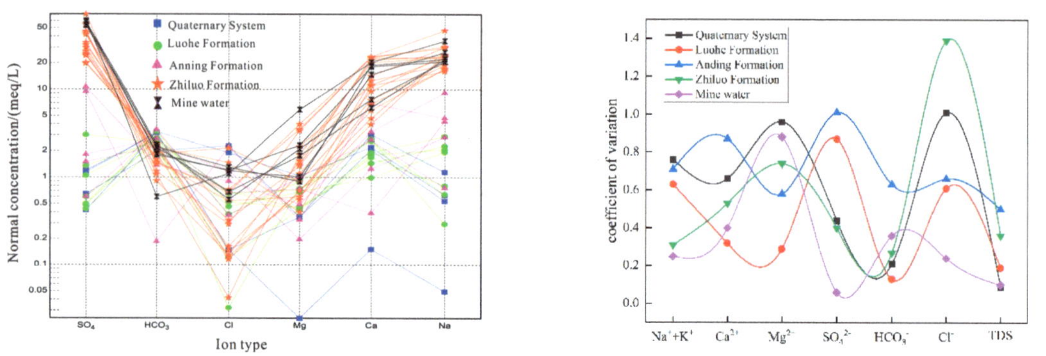

Figure 3. Schoeller diagram and variable coefficient diagram of ions. (**a**) Schoeller diagram; (**b**) Variable coefficient diagram of ions.

The variable coefficients of the ions or TDS concentrations were calculated for the quantitative dispersion of water quality (Figure 3b). The coefficient of variation positively correlated with the degree of data dispersion, indicating a higher fluctuation between the data points and suggesting a weak groundwater connection [29].

When combined with the Schoeller diagram, it became evident that SO_4^{2-} and Cl^- exhibit significant changes, resulting in different coefficients of variation. Conversely, HCO_3^- shows minimal change in the Schoeller diagram, leading to a low coefficient of variation. Generally speaking, Cl^- remains relatively stable in groundwater and, thus, does not display noticeable changes on the Schoeller diagram. However, Zhiluo Formation groundwater demonstrates a relatively large coefficient of variation and exhibits similar hydrochemical characteristics to mine water, indicating a close connection between them.

This is consistent with the conclusion that there is a close hydraulic connection between the Quaternary aquifer and the Luohe Formation, suggesting no effective water barrier between the Quaternary groundwater and the Cretaceous groundwater. The Luohe Formation aquifer groundwater, recharged by atmospheric precipitation, seeps vertically into the aquifer below the Zhiluo groundwater, which then recharges the mine water.

3.2. Characteristics of High-Salinity Mine Water

This chapter will specifically analyze the characteristics of mine water. According to Table 1, the TDS concentration of 38 water samples from the study area ranges from 216.99 to 5124.00 mg/L. Among them, there are 22 samples of high-salinity water with a TDS value exceeding 1000 mg/L, all of which are sourced from the Zhiluo groundwater and mine water.

The correlation diagram depicts the relationship between the TDS and ion concentrations of 22 high-salinity water samples, as illustrated (Figure 4). There is a positive correlation between TDS and ion concentrations in all mine water samples, except for HCO_3^-. In Zhiluo Formation water samples, there is a positive correlation between TDS and ion concentrations, except for HCO_3^- and Cl^-. The highest correlation is between TDS and SO_4^{2-}, with a correlation coefficient of 0.998. The correlation coefficients of Na^+, Mg^{2+}, and Ca^{2+} with TDS are 0.792, 0.788, and 0.759, respectively.

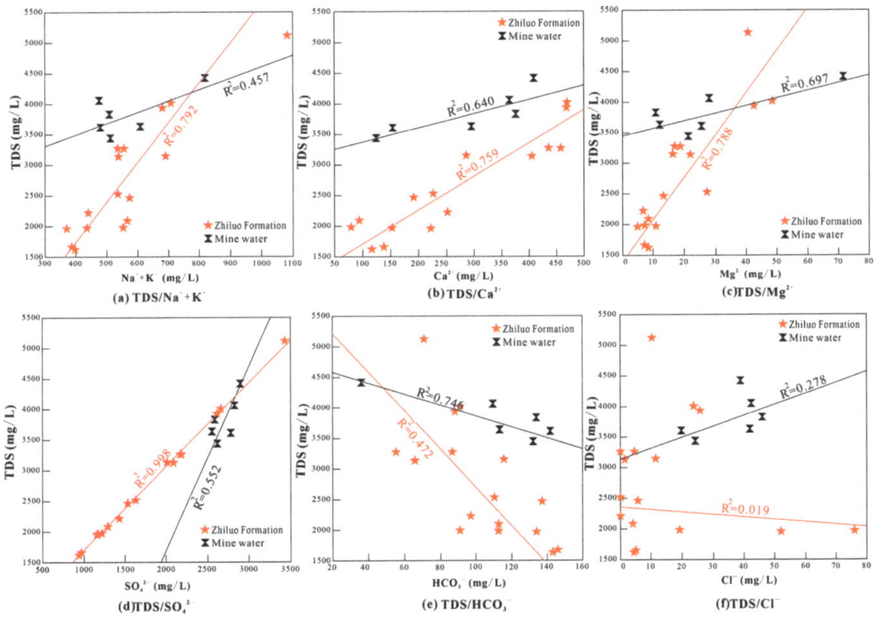

Figure 4. Correlation analysis chart of high-salinity water samples.

The results indicate that mine water and Zhiluo Formation groundwater have significant salinity, particularly in 22 high-salinity samples with TDS values exceeding 1000 mg/L. This high salinity is primarily contributed to by the Zhiluo Formation and mine water. The strong positive correlation between TDS and most ions (except HCO_3^-) in mine water samples and between TDS and most ions (except HCO_3^- and Cl^-) in Zhiluo Formation water samples highlights the influence of specific ions on overall salinity.

The exceptionally high correlation between TDS and SO_4^{2-} (0.998) suggests that sulfate ions play a major role in the salinity of these water samples. This is further supported by the notable correlation coefficients of Na^+ (0.792), Mg^{2+} (0.788), and Ca^{2+} (0.759) with TDSs in mine water samples. These findings indicate that sulfate, sodium, magnesium, and calcium are significant contributors to the high TDS levels observed.

Although the overall correlation between mine water and the Zhiluo Formation aquifer is relatively weak, the consistency in the composition of major ions indicates a hydrological connection between them. The correlation coefficients of Mg^{2+} (0.697) and Ca^{2+} (0.640) with TDS in the Zhiluo Formation aquifer further emphasize the importance of these ions in influencing the salinity.

In summary, the analysis reveals a clear pattern of ion contribution to the salinity in both mine water and Zhiluo Formation groundwater. The strong correlations between TDSs and specific ions, particularly SO_4^{2-}, Na^+, Mg^{2+}, and Ca^{2+}, suggest that these ions are key factors in determining salinity levels. The weak yet consistent overall correlation between the mine water and Zhiluo Formation aquifer composition suggests a significant hydrological interaction despite the relatively weak statistical correlation.

3.3. Formation of High-Salinity Mine Water

3.3.1. The Primary Source of Mine Water

The δD and $\delta^{18}O$ values in the water remain largely unaffected by water–rock interactions, enabling the effective determination of distinct water sources and their origins [29]. The δD and $\delta^{18}O$ isotopic values of Quaternary groundwater, the upper section of Zhiluo Formation groundwater, and the lower section of Zhiluo Formation groundwater and mine water are shown in Table 2. The stable isotope δD values of Zhiluo groundwater, Quaternary groundwater, and mine water exhibit a range from −83.6‰ to −56.3‰, while $\delta^{18}O$ values range from −10.8‰ to −7.7‰. According to the literature [30], if $\delta D < -400‰$ or $\delta D > 10‰$, and $-60‰ < \delta^{18}O < 0‰$, atmospheric precipitation is identified as the primary source.

Table 2. The average values of δD and $\delta^{18}O$ for various types of water samples from the study area.

Serial Number	Aquifer Position	Average Value of δD/‰	Average Value of $\delta^{18}O$/‰
1	Quaternary groundwater	−56.3	−7.7
2	Upper section of Zhiluo Formation	−80.8	−10.3
3	Lower section of Zhiluo Formation	−83.6	−10.8
4	Mine water	−81.9	−10.6

The average values of $\delta^{18}O$ and δD of Quaternary groundwater, upper and lower sections of the Zhiluo Formation groundwater, as well as mine water, are plotted in a relationship diagram (Figure 5).

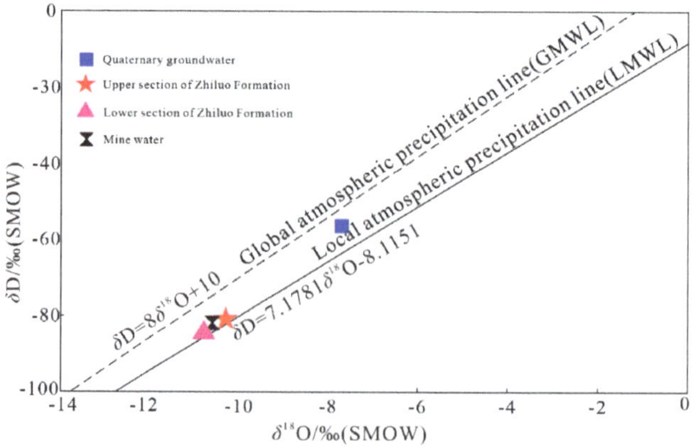

Figure 5. δD-$\delta^{18}O$ diagram illustrating water samples collected from the study area.

Due to the absence of local precipitation data, we rely on the equation of the atmospheric precipitation line in regions with analogous climatic conditions: $\delta D = 7.1781\ \delta^{18}O8.1151$ [30]. The δD and $\delta^{18}O$ values of Quaternary groundwater, upper Zhiluo Formation, lower Zhiluo Formation groundwater, and mine water closely align with the local atmospheric precipitation line, indicating that atmospheric precipitation serves as the primary source of recharge. The δD and $\delta^{18}O$ values of Quaternary water samples exhibit significant deviations from the local atmospheric precipitation line, which is primarily influenced by atmospheric rainfall, thereby distinguishing them from other water samples. According to d (deuterium surplus) = δD-$8\delta^{18}O$, the deuterium surplus of Zhiluo Formation and mine water ranges from 1.6 to 2.9 ‰, which is significantly below the threshold of 10‰, indicating a slow runoff rate and prolonged retention time within the aquifer. Comparing the average values of δD and $\delta^{18}O$ in Table 2, it has been determined that there is minimal disparity between the mean values of δD and $\delta^{18}O$ in mine water and those of water samples from the Zhiluo Formation. These findings indicate that the Zhiluo Formation aquifer water serves as the primary source of recharge for mine water, aligning with the aforementioned analysis results on water chemical composition.

3.3.2. Water–Rock Interaction

According to the proportional relationship between Na^+, Ca^{2+}, Cl^-, HCO_3^-, and TDS in water, the control factors of ion formation are determined by the Gibbs diagram [31,32]. The main control factors can be classified into three types: precipitation control type, rock-weathering control type, and concentration caused by evaporation. In the exploration of groundwater formation control, the graphical method has received a lot of attention from scholars at home and abroad. See Figure 6.

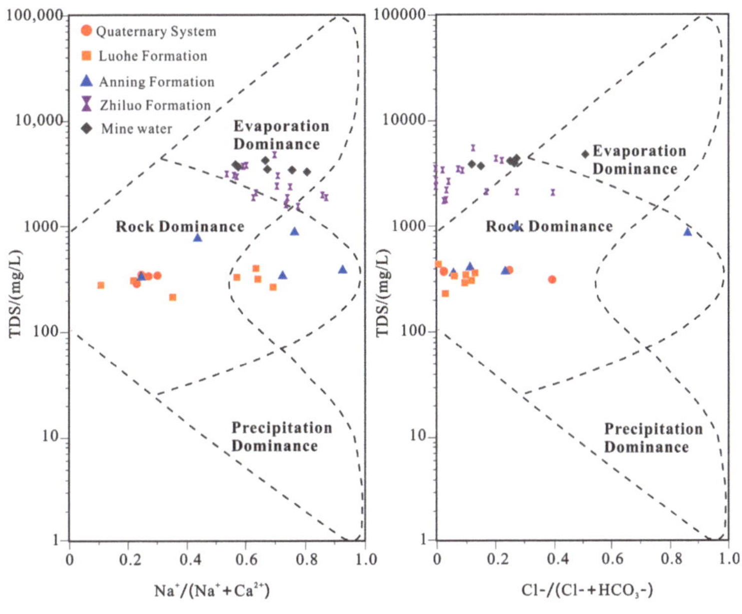

Figure 6. Gibbs diagram of water samples from study area.

The TDS of shallow groundwater is below 1000 $mg\cdot L^{-1}$, with $Na^+/(Na^+ + Ca^{2+})$ ratios ranging from 0.11 to 0.93 and $Cl^-/(Cl^- + HCO_3^-)$ ratios ranging from 0.03 to 0.87, all situated within the rock-weathering control area, indicating that the primary process responsible for the formation of shallow groundwater is rock-weathering control. The deep groundwater and mine water exhibit a TDS concentration exceeding 1000 $mg\cdot L^{-1}$, with $Na^+/(Na^+ + Ca^{2+})$ values ranging from 0.54 to 0.86 and $Cl^-/(Cl^- + HCO_3^-)$ ratios ranging from 0.00 to 0.52, predominantly below 0.3. The hydrogeology of the deep ZhiLuo formation and mine groundwater is primarily influenced by processes of concentration caused by evaporation. The chemical composition of deep groundwater is predominantly characterized as SO_4-Na type, exhibiting high mineralization and a relatively low concentration of Cl^-, resulting in the majority of water samples being distributed beyond the boundaries defined by the three control types. From the Gibbs diagram, it can be inferred that there are two main factors leading to the formation of highly mineralized mine water: the lixiviation and concentration caused by evaporation. The surrounding rock contains a large amount of soluble minerals. During the long-term contact between groundwater and the surrounding rock, the minerals gradually dissolve. Through relevant chemical reactions, the ion content in groundwater gradually increases. In addition, the research area is located in an arid and semi-arid region with strong evaporation. The evaporation and concentration of groundwater further increase the concentration of various ions in the water, leading to an increase in TDS values.

Groundwater formed under different geneses or conditions often has a large difference in the content between components. Therefore, such ratio coefficients can be used to determine the source of groundwater components [31]. Among them, the ratio relationship between Na^+ and Cl^- can be inferred as the source of Na^+. When the ratio of the two is about 1:1, it indicates that Na^+ in groundwater mainly comes from rock salt dissolution [32]. As shown in Figure 7, most of the samples from the Zhiluo Formation and mine water in the study area are located above the 1:1 line, and the equivalent concentration of Na^+ is significantly higher than that of Cl^-. This discovery indicates that Na^+ not only comes from the dissolution of rock salt but may also come from silica-aluminate minerals containing sodium or if there is a strong cation exchange interaction.

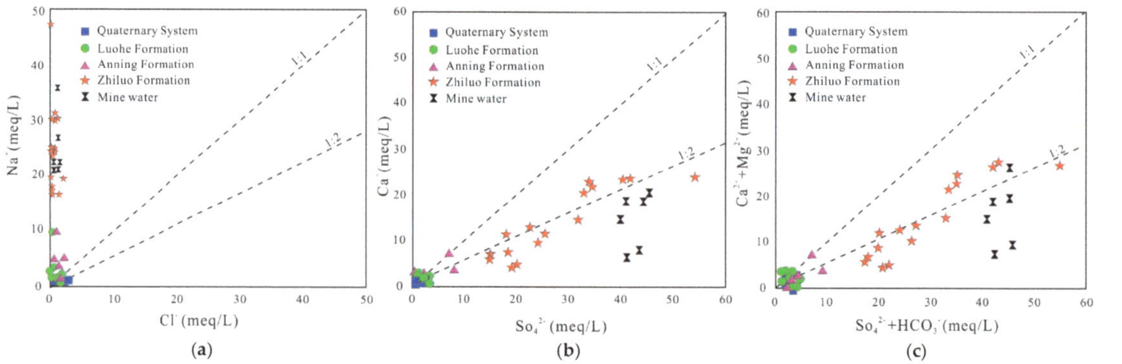

Figure 7. Ion ratio diagram. (**a**) Na^+ and Cl^- relational graph; (**b**) Ca^{2+} and SO_4^{2-} relational graph; (**c**) (Ca^{2+} +Mg^{2+}) and (SO_4^{2-}+HCO_3^-) relational graph.

The main sources of Mg^{2+} and Ca^{2+} can be determined by analyzing the ratio relationship between (Ca^{2+} + Mg^{2+}) and (HCO_3^- + SO_4^{2-}) [33]. The water samples from the Zhiluo Formation and mine in Figure 7b all fall below a 1:1 straight line, indicating that Mg^{2+} and Ca^{2+} in the groundwater mainly come from the weathering and dissolution of silicate and evaporite rocks rather than the weathering and dissolution of carbonate rocks. The approximate 1:1 ratio of Ca^{2+} to SO_4^{2-} in groundwater indicates that the primary source of these ions is the dissolution of gypsum. However, the samples from the Zhiluo Formation and mine water in the study area are generally positioned below the 1:1 line (Figure 7c). This suggests that other chemical processes, such as cation exchange, also influence the Ca^{2+} content in groundwater.

In summary, the dissolution of rock salt, gypsum, and other minerals provides a material foundation for the formation of high-salinity groundwater. The chemical equations for these processes are as follows:

1. Dissolution of rock salt:

$$NaCl\ (\text{rock salt}) \rightarrow Na^{+} + Cl- \tag{1}$$

2. Dissolution of sodium feldspar:

$$2NaAlSi_3O_8\ (\text{Albite}) + 2H_2CO_3 + 9H_2O \rightarrow Al_2Si_2O_5(OH)_4(\text{kaolinite}) + 2Na^{+} + 2\ HCO_3^{-} + 4H_4SiO_4 \tag{2}$$

3. Dissolution of gypsum:

$$CaSO_4\ (\text{gypsum}) \rightarrow Ca^{2+} + SO_4^{2-} \tag{3}$$

4. Solution of anorthite:

$$2CaAl_2Si_2O_8\ (\text{anorthite}) + 4CO_2 + 6H_2O \rightarrow 2Al_2Si_2O_5(OH)_4(\text{kaolinite}) + 2Ca^{2+} + 4HCO_3^{-} \tag{4}$$

3.3.3. Cation Exchange

The ratio of $(Ca^{2+} + Mg^{2+} - HCO_3^{-} - SO_4^{2-})/(Na^{+} - Cl^{-})$ serves as an indicator for assessing the impact of cation exchange on water quality [34–36]. When a linear relationship exists between the two parameters, with a slope of -1, it indicates that cation exchange is the primary process governing the hydrochemical evolution of water. As illustrated in Figure 8a, the ratio of $(Ca^{2+} + Mg^{2+} - HCO_3^{-} - SO_4^{2-})/(Na^{+} - Cl^{-})$ demonstrates a linear correlation with a slope of -0.65 ($R^2 = 0.65$), indicating the occurrence of cation exchange in groundwater during runoff. The analysis of major ion ratios and the aquifer rock composition mentioned earlier reveals that the concentrations of Ca^{2+} and Mg^{2+} are relatively low in both groundwater and surrounding rocks, thereby minimizing the impact of cation exchange.

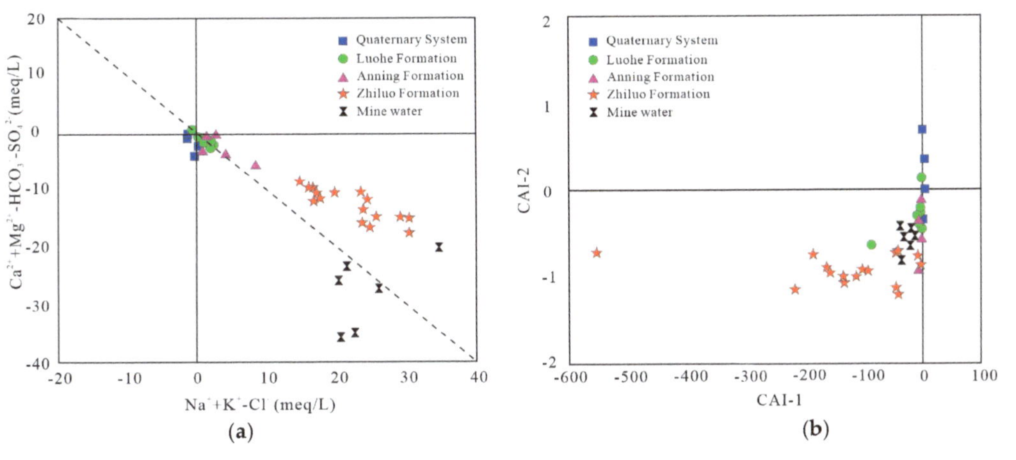

Figure 8. (**a**) $(Ca^{2+} + Mg^{2+} - HCO_3^{-} - SO_4^{2-})/(Na^{+} + K^{+} - Cl^{-})$ and (**b**) water Chlor-Alkali Index diagram.

The Chloro-Alkaline Index (CAI-1 and CAI-2) can be utilized to determine the process of ion exchange [37]. When the values of CAI-1 and CAI-2 are both less than 0, a reverse cation exchange occurs, where Ca^{2+} and Mg^{2+} undergo water exchange with Na^{+} and K^{+} on the mineral surface (Equation (7)). Otherwise, when the value is positive, a positive cation exchange occurs (Equation (8)), and the stronger the absolute value, the stronger the cation exchange [38].

The calculation formulas for CAI-1 and CAI-2 are as follows:

$$CAI\text{-}1 = [Cl^- - (Na^+ + K^+)]/Cl^- \quad (5)$$

$$CAI\text{-}2 = [Cl^- - (Na^+ + K^+)]/[HCO_{3^-} + SO_4^{2-} + CO_3^{2-} + NO_3^-) \quad (6)$$

The units of each index in the equation are meq/L.

The chemical reaction is represented as follows:

$$Ca^{2+}\ (Mg^{2+})(water) + Na^+(K^+)(rock) \rightarrow Na^+(K^+)(water) + Ca^{2+}(Mg^{2+})(rock) \quad (7)$$

$$Na^+(K^+)(water) + Ca^{2+}(Mg^{2+})(rock) \rightarrow Ca^{2+}(Mg^{2+})(water) + Na^+(K^+)(rock) \quad (8)$$

The Chloro-Alkaline Index diagrams depicting water samples from various aquifers are illustrated in Figure 8b. The Chlor-Alkali Index for the Anding Formation, Zhiluo Formation, and mine water is less than 0, as observed from Figure 8b, indicating that a positive cation exchange process has occurred. In Quaternary Formation groundwater, 75% and 14.3% are located in Quadrant I, indicating a process of reverse cation exchange, as well as the Luohe Formation. On the other hand, 25% and 85.7% are both located in Quadrant III, indicating a positive exchange of cations. The complexity of the cation exchange processes in the Quaternary System and the presence of both positive and reverse reactions indicate Luohe Formation groundwater. However, the dominant process in the Zhiluo Formation and mine water is positive cation exchange, resulting in an increased concentration of Na^+ in the water and the formation of highly mineralized groundwater.

4. Treatment and Utilization of High-Salinity Mine Water

High-salinity treatment technologies can be classified into three categories: chemical, thermal, and membrane separation methods [39–41]. Chemical methods primarily involve the use of reagents and ion exchange to remove salts from mine water through various chemical reactions. The reagent method typically uses limestone to reduce the high hardness levels in mine water. Ion exchange employs solid exchange agents to facilitate the ion exchange process, targeting ions such as calcium, magnesium, and chloride in mixed-type mine water. However, when the salinity exceeds 500 mg/L, the cost of this method increases significantly. Thermal methods use a heat source as the driving force for the separation of water and salts, thereby achieving demineralization. This method is suitable for mine water with total dissolved solids (TDSs) exceeding 3000 mg/L. Membrane Separation Demineralization Technology includes electrodialysis and reverse osmosis. Electrodialysis is suitable for treating mine water with a salt content between 500 mg/L and 4000 mg/L, whereas reverse osmosis is suitable for treating mine water with a salt content below 4000 mg/L and up to 10,000 mg/L [39,41].

Therefore, we chose reverse osmosis technology as the treatment method, which not only effectively reduced the salt content in water but, when combined with nanofiltration technology, also enhanced the purity of crystalline salt produced through evaporation crystallization.

Desalinated mine water can be used for industrial water supply for surrounding enterprises or for irrigation in nearby farmland, the domestic water needs of surrounding enterprises or residents, and ecological and environmental construction in mining areas [42,43]. This not only significantly reduces the discharge of mine water but also effectively alleviates the water shortage problem in mining areas. This helps to resolve the contradiction between water resource supply and demand in coal mining regions, achieving a harmonious balance between coal resource development and water resource protection and utilization.

5. Conclusions

Based on the comprehensive analysis of 38 groundwater and mine water samples using Piper diagrams, Gibbs diagrams, isotope analyses, and ion ratio coefficients, several key conclusions can be drawn. First, it was observed that the TDS concentration increases with depth, with the average TDS concentrations of Quaternary, Luohe, and Anding groundwater being 336.87 mg/L, 308.67 mg/L, and 556.29 mg/L, respectively. In stark contrast, the TDS concentrations of Zhiluo groundwater and mine water are significantly higher, at 2768.57 mg/L and 3826.40 mg/L, respectively, classifying them as high-salinity water. This pattern indicates a clear trend where deeper groundwater layers exhibit higher salinity levels.

The hydrochemical analysis revealed distinct types of groundwater within the study area. Quaternary groundwater is predominantly of the HCO_3-Ca type, while the Luohe Formation presents both HCO_3-Ca and HCO3-Na types. Anding Group's groundwater is characterized by the HCO_3-SO_4-Na·Ca type, and the Zhiluo Formation, along with mine water, is primarily of the SO_4-Na type. These classifications reflect the varied mineral compositions and geochemical processes affecting each groundwater source.

Isotope analysis further highlighted that the δD and $\delta^{18}O$ values of all groundwater and mine water samples were close to the meteoric water line, suggesting that atmospheric precipitation is the primary recharge source. The high salinity of Zhiluo Formation groundwater, which is the main source of mine water, is closely related to the dissolution of rock salt and gypsum, as well as cation exchange processes. This finding underscores the significant role of geological formations in contributing to the salinity levels observed in mine water.

Given the high TDS concentration of mine water, ranging from 3435.4 mg/L to 4414.3 mg/L, the combined treatment process of reverse osmosis is recommended for effective salt removal. The treated mine water can be repurposed for various uses, including industrial, domestic, and ecological applications. This approach not only aids in resource utilization but also supports the goal of achieving the zero discharge of mine water in Northwest China.

In summary, this study offers valuable insights into the formation mechanisms and treatment strategies for high-salinity mine water. While the findings provide a robust foundation for managing mine water resources, there are inherent limitations due to the availability and quality of data. Potential gaps in long-term hydrogeochemical monitoring and the precision of analytical methods in detecting trace elements at extremely low concentrations may affect the comprehensiveness of the conclusions. Additionally, although the sample size was carefully selected, it may not fully capture the variability within the larger aquifer system. These limitations highlight the need for continued research and more extensive data collection to refine our understanding and management of high-salinity mine water.

Author Contributions: Conceptualization, J.Y. and W.Z.; methodology, W.Z.; software, W.Z. and J.Y.; validation, J.Y. and W.Z.; formal analysis, J.Y., W.Z. and X.L.; investigation, X.L. and F.X.; data curation, X.L. and F.X.; writing—original draft preparation, J.Y. and W.Z.; writing—review and editing, J.Y. and W.Z.; visualization, W.Z.; supervision, X.L. and F.X. All authors have read and agreed to the published version of the manuscript.

Funding: This research was financially supported by the Natural Science Foundation of Shaanxi province (2023-JC-QN-0291) and National Key Research and Development Program of China (2016YFC0501105).

Institutional Review Board Statement: Not applicable.

Informed Consent Statement: Not applicable.

Data Availability Statement: The data presented in this study are available on request from the corresponding author.

Conflicts of Interest: Authors Xiangyang Liang and Feng Xu were employed by the company Xi'an Research Institute of China Coal Technology & Engineering, (Group), Corp. The remaining authors declare that the research was conducted in the absence of any commercial or financial relationships that could be construed as a potential conflict of interest.

References

1. Al-Aizari, H.S.; Aslaou, F.; Al-Aizari, A.R.; Al-Odayni, A.-B.; Al-Aizari, A.-J.M. Evaluation of Groundwater Quality and Contamination Using the Groundwater Pollution Index (GPI), Nitrate Pollution Index (NPI), and GIS. *Water* **2023**, *15*, 3701. [CrossRef]
2. Boretti, A.; Rosa, L. Reassessing the projections of the world water development report. *NPJ Clean Water* **2019**, *2*, 15. [CrossRef]
3. Zhang, Q.; Miao, L.; Wang, H.; Hou, J.; Li, Y. How Rapid Urbanization Drives Deteriorating Groundwater Quality in a Provincial Capital of China. *Pol. J. Environ. Stud.* **2020**, *29*, 441–450. [CrossRef] [PubMed]
4. Mao, J.; Tong, H. Coal resources, production, and use in China. *Coal Handb.* **2023**, *2*, 431–454. [CrossRef]
5. Ren, S.; Jiao, X.; Zheng, D.; Zhang, Y.; Xie, H.; Guo, Z. Demand and fluctuation range of China's coal production under the dual carbon target. *Energy Rep.* **2024**, *11*, 3267–3282. [CrossRef]
6. Gu, H.; Lai, X.; Tao, M.; Momeni, A.; Zhang, Q.I. Dynamic mechanical mechanism and optimization approach of roadway surrounding coal water infusion for dynamic disaster prevention. *Measurement* **2023**, *223*, 113639. [CrossRef]
7. Qian, X.; Wang, D.; Wang, J.; Chen, S. Resource curse, environmental regulation and transformation of coal-mining cities in China. *Resour. Policy* **2021**, *74*, 101447. [CrossRef]
8. Singh, R.; Venkatesh, A.S.; Syed, T.H.; Reddy, A.G.S.; Kumar, M.; Kurakalva, R.M. Assessment of potentially toxic trace elements contamination in groundwater resources of the coal mining area of the Korba Coalfield, Central India. *Environ. Earth Sci.* **2017**, *76*, 566. [CrossRef]
9. Wang, H.; Dong, S.; Shang, H.; Wang, T.; Yang, J.; Zhao, C.; Zhang, Q.; Zhou, Z.; Liu, J.; Hou, Y. Domestic and foreign progress of mine water treatment and resource utilization. *Coal Geol. Explor.* **2023**, *51*, 222–236. [CrossRef]
10. Feng, H.; Zhou, J.; Chai, B.; Zhou, A.; Li, J.; Zhu, H.; Su, D. Groundwater environmental risk assessment of abandoned coal mine in each phase of the mine life cycle: A case study of Hongshan coal mine, North China. *Environ. Sci. Pollut. Res.* **2020**, *27*, 42001–42021. [CrossRef]
11. Zhao, Q.; Guo, F.; Zhang, Y.; Ma, S.; Jia, X.; Meng, W. How sulfate-rich mine drainage affected aquatic ecosystem degradation in northeastern China, and potential ecological risk. *Sci. Total Environ.* **2017**, *609*, 1093–1102. [CrossRef]
12. Xu, Y.; Ma, L.; Khan, N.M. Prediction and maintenance of water resources carrying capacity in mining area—A case study in the yu-shen mining area. *Sustainability* **2020**, *12*, 7782. [CrossRef]
13. Wang, X.; Zhang, J.; Elmahdi, A.; Shamsuddin, S.; Gao, J. A water resources assessment framework for management strategies of large coal-power bases development in China. *Mitig. Adapt. Strateg. Glob. Change* **2023**, *28*, 35. [CrossRef]
14. Nagaraju, A.; Sunil Kumar, K.; Thejaswi, A. Assessment of groundwater quality for irrigation: A case study from Bandalamottu lead mining area, Guntur District, Andhra Pradesh, South India. *Appl. Water Sci.* **2014**, *4*, 385–396. [CrossRef]
15. Lautz, L.K.; Hoke, G.D.; Lu, Z.; Siegel, D.I.; Christian, K.; Kessler, J.D.; Teale, N.G. Using discriminant analysis to determine sources of salinity in shallow groundwater prior to hydraulic fracturing. *Environ. Sci. Technol.* **2014**, *48*, 9061–9069. [CrossRef]
16. Wen, X.; Wu, Y.; Su, J.; Zhang, Y.; Liu, F. Hydrochemical characteristics and salinity of groundwater in the Ejina Basin, Northwestern China. *Environ. Geol.* **2005**, *48*, 665–675. [CrossRef]
17. Sun, Y.; Mao, X.; Yang, X.; Dong, L.; Tang, M. Variation of groundwater salinity and its influence on crops in irrigation area of Northwest China. *Trans. Chin. Soc. Agric. Eng.* **2010**, *26*, 103–108.
18. Wang, P.; Yang, L.; Lin, X.; Yang, X. Distribution characteristics and formation of high mineralized saline groundwater in Hetao Plain, Inner Mongolia. *Yangtze River* **2018**, *49*, 44–50. [CrossRef]
19. Lv, X.; Shao, J.; Liu, J.; Wen, J.; Sun, J. Distribution characteristics and origin of total dissolved solids in groundwater under Lanzhou City. *J. Arid Land Resour. Environ.* **2013**, *27*, 23–27. [CrossRef]
20. Jin, D.; Wang, T.; Zhao, B.; Li, D.; Zhou, Z.; Shang, H. Distribution characteristics and formation mechanism of high salinity groundwater in northeast Ningdong Coalfield. *Coal Geol. Explor.* **2022**, *50*, 118–127. [CrossRef]
21. Zhang, Y.; Xu, Z.; Zhang, L.; Lv, W.; Yuan, H.; Zhou, L.; Gao, Y.; Zhu, L. Hydrochemical characteristics and genetic mechanism of high TDS groundwater in Xinjulong Coal Mine. *Coal Geol. Explor.* **2021**, *49*, 52–62. [CrossRef]
22. Hou, G.; Gao, M.; Ye, S.; Zhao, G. Source of salt and the salinization process of shallow groundwater in the Yellow River Delta. *Earth Sci. Front.* **2022**, *29*, 145–154. [CrossRef]
23. He, M.; Zhang, B.; Xia, W.; Cui, X.; Wang, Z. Hydrochemical Characteristics and Analysis of the Qilihai Wetland, Tianjin. *Environ. Sci.* **2021**, *42*, 776–785. [CrossRef]
24. Wu, D.; Yao, Z.; Jia, F.; Wei, X.; Sun, H.; Mao, Q.; Xie, N. Hydro—Geochemical characteristics and genetic analysis of groundwater in Ha-mi basin, Xinjiang. *J. Arid Land Resour. Environ.* **2020**, *34*, 133–141. [CrossRef]
25. Fu, C.; Liu, C. Hydrochemical characteristics and formation mechanism of saline lakes in Hoh Xil region. *Yangtze River* **2022**, *53*, 36–41. [CrossRef]
26. He, X.; Hu, D.; Hu, Z.; Wang, P. Research on technology for high mineralized mine water treatment. *Coal Sci. Technol.* **2002**, *30*, 38–41. [CrossRef]
27. Wang, J.; Meng, X.; Chen, L.; Zhao, J. Advances in Causes of Brackish Water and Desalination Technology. *Gansu Agric. Sci. Technol.* **2010**, *7*, 39–42. [CrossRef]

28. Cates, D.A.; Knox, R.C.; Sabatini, D.A. The Impact of Ion Exchange Processes on Subsurface Brine Transport as Observed on Piper Diagrams. *Groundwater* **1996**, *34*, 532–544. [CrossRef]
29. Zhou, J.; Liu, G.; Meng, Y.; Xia, C.; Chen, K. Characteristics of δ 18O and δ 2H and their implication for the interaction between precipitation, groundwater and river water in the upper River Tuojiang, Southwest China. *Water Pract. Technol.* **2021**, *16*, 226–246. [CrossRef]
30. Jin, J.; Wang, Z.; Zhao, Y.; Ding, H.; Zhang, J. Delineation of Hydrochemical Characteristics and Tracing Nitrate Contamination of Groundwater Based on Hydrochemical Methods and Isotope Techniques in the Northern Huangqihai Basin, China. *Water* **2022**, *14*, 3168. [CrossRef]
31. Tang, R.; Dong, S.; Zhang, M.; Zhou, Z.; Zhang, C.; Li, P.; Bai, M. Hydrochemical Characteristics and Water Quality of Shallow Groundwater in Desert Area of Kunyu City, Southern Margin of Tarim Basin, China. *Water* **2023**, *15*, 1563. [CrossRef]
32. Marandi, A.; Shand, P. Groundwater Chemistry and the Gibbs Diagram. *Appl. Geochem.* **2018**, *97*, 209–212. [CrossRef]
33. Sunkari, E.D.; Abu, M.; Zango, M.S. Geochemical evolution and tracing of groundwater salinization using different ionic ratios, multivariate statistical and geochemical modeling approaches in a typical semi-arid basin. *J. Contam. Hydrol.* **2021**, *236*, 103742. [CrossRef]
34. Li, P.; Wu, J.; Qian, H. Hydrochemical appraisal of groundwater quality for drinking and irrigation purposes and the major influencing factors: A case study in and around Hua County, China. *Arab. J. Geosci.* **2016**, *9*, 15. [CrossRef]
35. Zhang, C.; Luo, B.; Xu, Z.; Sun, Y.; Feng, L. Research on the Capacity of Underground Reservoirs in Coal Mines to Protect the Groundwater Resources: A Case of Zhangshuanglou Coal Mine in Xuzhou, China. *Water* **2023**, *15*, 1468. [CrossRef]
36. Rao, N.S.; Sunitha, B.; Sun, L.; Spandana, B.D.; Chaudhary, M. Mechanisms controlling groundwater chemistry and assessment of potential health risk: A case study from South India. *Geochemistry* **2020**, *80*, 125568. [CrossRef]
37. Sharma, N.; Vaid, U.; Sharma, S.K. Assessment of groundwater quality for drinking and irrigation purpose using hydrochemical studies in Dera Bassi town and its surrounding agricultural area of Dera Bassi Tehsil of Punjab, India. *SN Appl. Sci.* **2021**, *3*, 245. [CrossRef]
38. Zhao, X.; Xu, Z.; Sun, Y. Mechanism of Changes in goaf water hydrogeochemistry: A case study of the menkeqing coal mine. *Int. J. Environ. Res. Public Health* **2022**, *20*, 536. [CrossRef]
39. Abdykadyrov, A.; Abdullayev, S.; Tashtay, Y.; Zhunussov, K.; Marxuly, S. Purification of Surface Water by Using the Corona Discharge Method. *Min. Miner. Depos.* **2024**, *18*, 125–137. [CrossRef]
40. Fitzsimons, E.; Warren, P. Desalination Investment for Copper Mining: Barriers and Opportunities in Chile. *Extr. Ind. Soc.* **2024**, *17*, 101449. [CrossRef]
41. Bondarenko, V.; Salieiev, I.; Kovalevska, I.; Chervatiuk, V.; Malashkevych, D.; Shyshov, M.; Chernyak, V. A New Concept for Complex Mining of Mineral Raw Material Resources from DTEK Coal Mines Based on Sustainable Development and ESG Strategy. *Min. Miner. Depos.* **2023**, *17*, 1–16. [CrossRef]
42. Zhang, S.; Wang, H.; He, X.; Guo, S.; Xia, Y.; Zhou, Y.; Liu, K.; Yang, S. Research progress, problems and prospects of mine water treatment technology and resource utilization in China. *Crit. Rev. Environ. Sci. Technol.* **2019**, *50*, 331–383. [CrossRef]
43. Ge, G.; Wu, Y.; Zhang, Q. Research on technology and process for moderate desalination of high-salinity mine water by nanofiltration. *Coal Sci. Technol.* **2021**, *49*, 208–214.

water

MDPI

Article

Extension Mechanism of Water-Conducting Cracks in the Thick and Hard Overlying Strata of Coal Mining Face

Dong Wei [1,2], Helong Gu [2,*], Chungang Wang [1], Hao Wang [2], Haoyu Zhu [2] and Yuyang Guo [2]

[1] Shanxi Yanchang Petroleum and Mining Co., Ltd., Yulin 710054, China; weidong6818@sxycpc.com (D.W.)
[2] College of Energy Engineering, Xi'an University of Science and Technology, Xi'an 710054, China; wanghao94100@163.com (H.W.); gyy18220222665@163.com (Y.G.)
* Correspondence: helonggu@xust.edu.cn

Abstract: It is of great significance for coal safety production and water resource protection in the Yuheng mining area to master the evolution law of water-conducting fractures under the condition of thick and hard overburden. This research focuses on the 2102 fully mechanized mining face in the Balasu Coal Mine as the research background. The fracture evolution and strata movement characteristics in thick and hard overlying strata are simulated and analyzed by combining numerical simulation with physical simulation, and the formation mechanism of a water-conducting fracture in the overlying strata is revealed and verified by field measurements of the development height of "two zones". The results show that the anisotropy of fracture propagation in low-position overlying strata is high, and the fracture propagation in high-position overlying strata is mainly vertical, which indicates characteristics of leapfrog development. The number and development height of fractures undergo the change–growth process of "slow–rapid–uniform". Multiple rock strata together form a complex force chain network with multiple strong chain arches. The local stress concentration leads to the initiation of micro-cracks in contact fractures, and the cracks gradually penetrate from bottom to top and then the strong chain arches are broken. The water-conducting cracks in overlying strata show a dynamic expansion process of "local micro-cracks–jumping cracks–through cracks–water-conducting cracks". The fracture between the caving zone and fracture zone presents obvious layered characteristics, the overall shape of the water-conducting fracture zone is "saddle-shaped", and the maximum development height lags behind the coal mining face by about 180 m. Through the observation of water injection leakage and borehole TV observation of three boreholes under underground construction, combined with the results of water pressure tests, it is comprehensively determined that the height of the water-conducting fracture zone is 103.68~107.58, and the fracture–production ratio is 31.42~32.60, which is basically consistent with the results of numerical simulation and physical simulation. This research provides theoretical guidance and a scientific basis for coal mine water disaster prevention under similar geological conditions.

Keywords: thick hard overburden; fracture evolution; height of water-conducting fracture; strata movement; particle flow

Citation: Wei, D.; Gu, H.; Wang, C.; Wang, H.; Zhu, H.; Guo, Y. Extension Mechanism of Water-Conducting Cracks in the Thick and Hard Overlying Strata of Coal Mining Face. *Water* **2024**, *16*, 1883. https://doi.org/10.3390/w16131883

Academic Editor: Fernando António Leal Pacheco

Received: 28 May 2024
Revised: 25 June 2024
Accepted: 26 June 2024
Published: 1 July 2024

1. Introduction

Mine water hazards, as one of the most dangerous disasters in coal mining, not only threaten the safety of miners but can also lead to severe economic losses and environmental damage [1,2]. Despite extensive research into the mechanisms and control technologies for mine water hazards over the past decades, the diversity and complexity of geological conditions, especially in regions like the Yuheng mining area in China, still pose significant challenges to water hazard prevention. The coal mining areas in Western China, particularly the Yuheng mining area, are critically important for the study of mine water hazards due to their complex hydrogeological characteristics and frequent human activities. In these areas, water hazards are influenced not only by natural geological factors but also by the mining methods, progress, and waterproofing measures employed within the mines. Current research and practice indicate that traditional methods for controlling water hazards are often ineffective under these complex conditions; hence, it is imperative to conduct more in-depth research to uncover the underlying causes and mechanisms of mine water hazards, particularly the evolution of water-conducting fractures in the overburden affected by mining activities. Understanding these fundamental scientific issues can provide a theoretical and technical basis for devising more effective prevention measures and response strategies, which are crucial for ensuring the safe production and sustainable development of mining areas [3].

Under the conditions of hard roof mining in a thick coal seam, a coal seam mining crack is the main channel of water inrush in a coal mining face. Engineering practice and research have shown that after mining, the overburden can be divided into caving zones, fracture zones, and bending subsidence zones, where the caving and fracture zones together form the water-conducting channels in the mining area [4–6]. Many scholars have studied the development characteristics and height calculation methods of water-conducting fracture zones in thick weakly cemented overburden. Y. Chen and S. Zhu [7] conducted specific studies on the "two soft and one hard" coal seams, determining the heights of the "two zones" in the overburden under these conditions. B. Hebblewhite [8] established a primary predictive model for the impact of rock layer displacement and fracturing on groundwater above longwall faces in the southern coalfields of New South Wales. Q. Ye et al. [9] experimentally determined the fracture evolution patterns in overburden under mining conditions. G. Wang et al. [10] studied the height of fracture zones in the overburden under mining conditions through physical simulation experiments. Xu et al. [11] used hierarchical analysis and fuzzy comprehensive evaluation methods to assess and zone the water resource carrying capacity in the Yushen mining area. Liu et al. [12], based on 203 borehole data from the Yushenfu mining area, developed a zoning method for the impact of mining on groundwater resources, categorizing the impact levels as severe, moderate, minor, and no leakage. Singh [13] and Coe [14] highlighted the importance of aquiclude rock groups located between water bodies and mined-out areas, concluding the impact of longwall mining on overlying strata. Booth [15] noted that changes in hydraulic parameters after mining primarily depend on the stratigraphic relationship between the rock layers and the mined-out areas. The complexity of these hydraulic changes is a significant factor in causing connectivity between aquifers and mined-out areas. In other research aspects of working face mining under the conditions of a hard roof in a thick coal seam, Jia [16] studied the stress distribution of an irregular island working face affected by a hard roof in a thick coal seam and the pressure relief control of surrounding rock in a roadway. Zhang [17] optimized the mixed materials and analyzed the mechanical failure of the filling materials effectively controlled by mining in his working face. Huang [18] analyzed the fracture instability characteristics and fracture mechanism of overlying strata in a large mining height section.

This paper takes the 2102 working face of the Balasu Coal Mine in the Yuheng mining area as the background, applying an integrated method of numerical and physical simulation to analyze the characteristics and expansion patterns of overburden fractures. It reveals the mechanisms of water-conducting fracture channels formed under the force chain action induced by mining activities. This study's accuracy is validated through field measurements, aiming to provide a basis for safe coal mining and water resource protection in the Yuheng mining area.

2. Engineering Background

2.1. General Siruation of Work Face

The aforementioned studies provide substantial theoretical and technical guidance for addressing the intertwined challenges of coal mining and water resource protection in the Yuheng mining area. However, research has primarily focused on the morphology, height evolution, and sensitivity of controlling factors of water-conducting fractures, with minimal exploration into the mechanisms of overburden fracture expansion and the cross-scale formation processes of water-conducting fractures. Therefore, this paper uses the Balasu Coal Mine's 2102 coal mining face in the Yuheng mining area as a case study, where the coal seam thickness is 3.64 m. During this process, both numerical and physical simulation methods were employed to analyze the patterns of fracture expansion in thick, hard overburden and the formation mechanisms of water-conducting fracture channels under mining influences, verified by field measurements to provide a foundation for safe coal mining and water resource protection in the Yuheng mining area.

2.2. Mine Waterfiling Factors

According to the geological and hydrogeological data, the Balasu Coal Mine features a complex hydrogeological type. The aquifers in the mine, from top to bottom, are as follows: Quaternary unconsolidated rock pore and fracture aquifers, Cretaceous Luohe Formation sandstone pore and fracture aquifers, Jurassic Zhiruo Group, and Yan'an Formation clastic rock fracture-confined aquifers, as shown in Figure 1. The Quaternary and Cretaceous aquifers, which indirectly recharge the aquifer above Coal Seam No. 2, have a minor impact on coal mining. The Jurassic Zhiruo Group fracture-confined aquifers and the Yan'an Formation sandstone fracture aquifers, which directly recharge the aquifer above Coal Seam No. 2, allow groundwater to flow into the mine along the water-conducting fracture zone. These aquifers have weak water richness and low recharge intensity. When aquifers with moderate or higher water richness or other water bodies pose a threat to the coal seam roof, it is essential to measure the development height of the collapse zone and the water-conducting fracture zone. This research focuses on the development height and characteristics of the mining-induced water-conducting fracture zone in Coal Seam No. 2 of the Balasu Coal Mine.

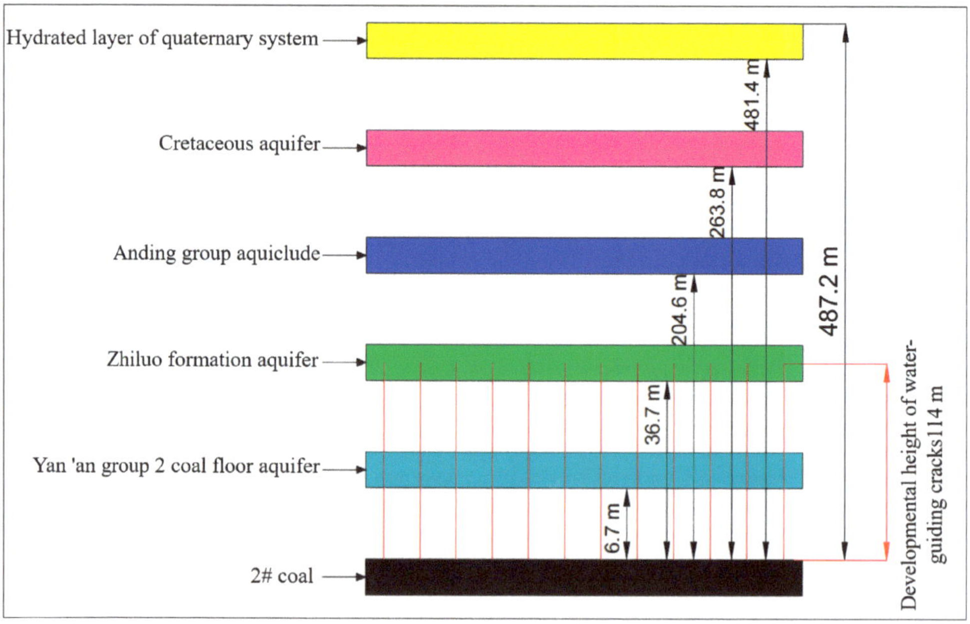

Figure 1. Strata lithology and aquifer distribution of 2102 coal mining face.

Considering the geological conditions of the Balasu Coal Mine, the main source of water flooding the mine's deposits is groundwater. The principal aquifers in the area are the Cretaceous Luohe sandstone pore and fracture water and the Quaternary Sarawusu

Formation pore aquifers, which are widely distributed throughout the mine with substantial thickness and medium-to-strong water richness. Both the main and auxiliary shafts must penetrate these layers during development, posing the most significant water hazard during the construction of the Balasu Coal Mine. Seam No. 2 is the first coal seam to be mined in the various mines of the Yuheng North area, and its aquifer represents the unique hydrogeological conditions of the Yuheng North area. From what has been revealed in surrounding mines, both the Xiaojihan and Yuan Datian mines experienced sudden water inrushes when Seam No. 2 was exposed. Currently, these two mines still experience considerable water inflow during the advancement through Coal Seam No. 2, with the primary source of mine water inflow being from the coal seam during the driving phase, while the water output from the sandstone roof during the retreat mining phase is relatively smaller. According to hydrological survey results, the aquifers in the Yan'an Formation above Coal Seam No. 2 predominantly consist of fine to medium-grained feldspathic sandstone, which is often calcite cemented and dense, with underdeveloped fractures and poor permeability. The water table depth ranges from 22.91 to 31.28 m, with water table elevations between +1153.96 and +1184.95 m, borehole water inflow rates of 49.85 to 68.59 m^3/day, specific inflow rates of 0.0091 to 0.0094 L/s·m, and permeability coefficients of 0.043 to 0.062 m/day. The aquifer has weak water richness; its hydrochemical type is predominantly SO4-Na, with pH values ranging from 8.38 to 8.88, indicating poor water quality. Consequently, the Balasu Coal Mine faces significant water hazard issues associated with highly water-rich aquifers during the exposure of Coal Seam No. 2 and the advancement of mine tunnels.

3. Evolution and Expansion Mechanism of Water-Conducting Fracture in Thick and Hard Overburden Rock

During the mining-induced instability and fracturing process, the overburden exhibits typical characteristics of discontinuous failure. The PFC2D particle flow discrete element software effectively overcomes the continuity assumptions of traditional continuous media, simulating the mechanical problems of discontinuous media accurately. It captures the discontinuous phenomena such as cracking, fracturing, and breaking of the coal rock mass, describing the dynamic evolution of microcrack generation, extension, and expansion during rock mass failure from a micro-viewpoint [19,20].

3.1. Numerical Modeling

In the PFC model, particle motion follows Newton's second law, with interparticle forces transmitted via contact. The fundamental contact models are divided into the bonded contact model and the parallel bond model, the latter referred to here as the PB Model, which uses line contacts to transmit both forces and moments, making it particularly suitable for the mechanical analysis of rock-like materials [21,22]. The propagation of microcracks easily leads to the generation of acoustic emission/microseismic energy, and cracks penetrate and then form water channels [23,24]. As shown in Figure 2, in PFC, cracks form when the relative movement of particles alters the stress environment of the bonds, leading to fracture when the tensile and shear stresses exceed the normal and tangential bond strengths, resulting in tensile and shear cracks between particles. Before generating the model, it is necessary to scientifically convert the micro-parameters of particles and bonds to align with the real macroscopic mechanical parameters of the rock. The micro-parameters needed for calibrating the PB Model include contact modulus E^*, stiffness ratio K^*, normal cohesive strength c, tangential cohesive strength t, angle of internal friction φ, and friction coefficient μ. As the PB Model degrades to a linear model after bond failure, the model's micro-parameters are simplified under the assumptions that: (1) the linear contact modulus between particles equals the parallel bond modulus; and (2) the ratio of tangential to normal stiffness between particles matches that of the parallel bonds. The micro-parameters for the PB Model are determined by trial and error, and calibration is considered successful if the deviation between the calculated rock stress–strain curve and laboratory values is within 5% [25–29]. The calibrated micro-parameters of the PB Model are listed in Table 1.

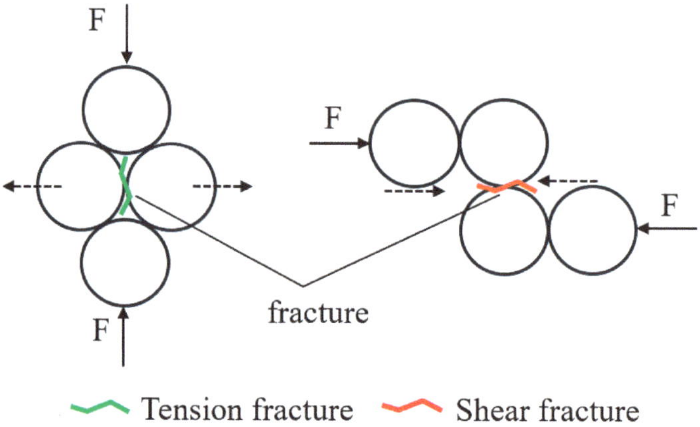

Figure 2. PFC crack propagation principle.

Table 1. Stratigraphic mechanical parameters of the PFC2D numerical model.

Lithologic Characters	E*/GPa	K*	c/MPa	t/MPa	φ/°	μ
Fine sandstone	10.5	1.6	12.8	8.5	39.4	0.7
Siltstone	8.8	1.5	11.7	7.4	38.2	0.8
Medium sandstone	7.4	1.7	6.2	5.6	35.2	0.7
Grit stone	8.2	1.7	9.9	7.2	39.5	0.6
Coal seam	2.8	1.9	3.2	2.1	29.8	0.4

For the geological conditions of Balasu Coal Mine's 2102 longwall face, a PFC2D numerical model for along-strike mining was constructed, measuring 420 m in length and 240 m in height. The model constrains the displacement at both sides and the bottom, with the top 250 m of overburden replaced by an equivalent load of 6.25 MPa, implemented via servo control of the top wall using Fish language. The model sets each stratum's sequence, lithology, and thickness according to actual borehole exploration results. Considering computational efficiency and effectiveness, particle radii were set between 0.5 and 0.7 m, generating 77,686 particles with 197,262 contacts. The model equilibrium condition was set to cease calculations when the maximum unbalanced force reached 1×10^{-5}. The actual advance rate of the 2202 working face was 8.0 m/day, with a single excavation step of 8 m, as shown in Figure 3. To represent the structural characteristics of bedding planes between different lithological layers, an unbonded Smooth Joint Model was implemented. After applying microparameters as shown in Table 1, the model reached equilibrium under a top overburden load of 6.25 MPa and gravitational force of 9.8 m/s^2. The computation at equilibrium took 34,685 steps. The initial numerical model is shown in Figure 3, with particle contact strengths between 6×10^6 and 26×10^6 N, and vertical stress increasing from 6.25 MPa at the top to 12.0 MPa.

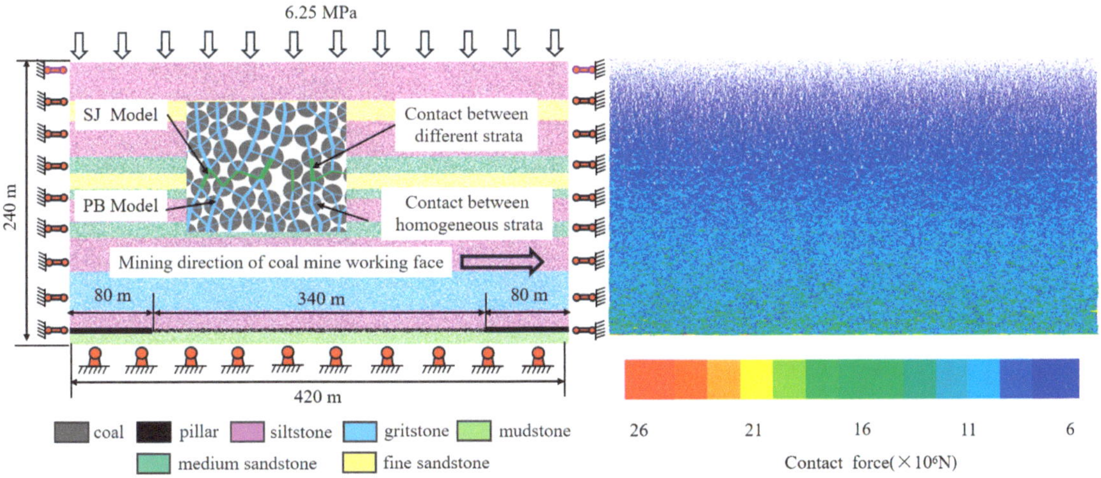

Figure 3. PFC numerical model construction.

3.2. Simulation Result Analysis

3.2.1. Overburden Rock Fracture Evolution Characteristics

As the coal mining face advances, the overburden fractures undergo an evolutionary process from "microcrack initiation to gradual coalescence to stable expansion", as depicted in Figure 4. At 48 m of advancement, vertical through-going fractures appear on both sides of the goaf in the immediate roof layer, with a relatively higher number of fractures near the cut-eye side. Locally, four microcracks emerge in the rock layers 10 to 19 m above the mine floor, where fracture spatial distribution is more dispersed and fewer in number. At 80 m of advancement, microcracks above the goaf vertically expand and gradually penetrate the rock layers. Concurrently, the direction of fracture expansion at the rock layer interfaces shifts to lateral expansion, producing larger-scale delamination fractures 9 m above the mine floor (due to the use of the Smooth Joint (SJ) Model at different rock layer interfaces, these delamination fractures cannot be displayed as red short lines in the software). At this stage, vertical through-going fractures develop up to 19 m above the mining area, with multiple sets of through-going fractures segmenting the rock layers into blocks. The face also enters the abutment pressure phase, with nine microcracks developing in higher rock layers (32~40 m above the mining area). At an advancement of 112 m, vertical through-going fractures continue to propagate into higher rock layers, with fracture development on the sides of the goaf noticeably lagging behind the central part, displaying a distribution characteristic of higher fractures in the center and lower on the sides. A new delamination fracture appears at 18 m above the mine floor, and microcracks develop locally in even higher overlying rock. When the face advances to 144 m, the fracture morphology in the overburden shows distinct stratification characteristics. While fractures in lower rock layers penetrate vertically, some also extend laterally, whereas fractures in upper rock layers primarily extend vertically and exhibit leapfrog development, i.e., multiple rock layers internally exhibit vertical fractures but are not completely interconnected. The delamination fractures previously observed at 9 m and 19 m above the original mine floor close under the compacting action of the overlying rock layers, and a new delamination fracture appears at 28 m above the mine floor. At a 176 m advance of the coal mining face, the stratification characteristics of the overburden fractures become more pronounced, with anisotropy in fracture development directions increasing in the lower rock layers, resulting in a complex fracture network. The higher rock layers continue to be dominated by vertical fractures. At this stage, the leapfrog vertical fractures on the cut-eye side and in the central part of the goaf gradually interconnect to form through-going vertical fractures, with limited vertical extension of fractures above the mining area, displaying a leapfrog vertical fracture pattern. As mining continues, the spatial distribution characteristics and development height of the fractures remain essentially unchanged, achieving the full mining effect, with overburden fractures periodically extending laterally as the coal mining face advances. Given that through-going vertical fractures serve as the primary channels for overburden water to enter the mining area, the height of these through-going vertical fractures can be defined as the height of the water-conducting fracture zone. During the advancement from 176 m to 340 m, the height of this fracture zone remains essentially unchanged, with the maximum development height of the water-conducting fractures being 108 m.

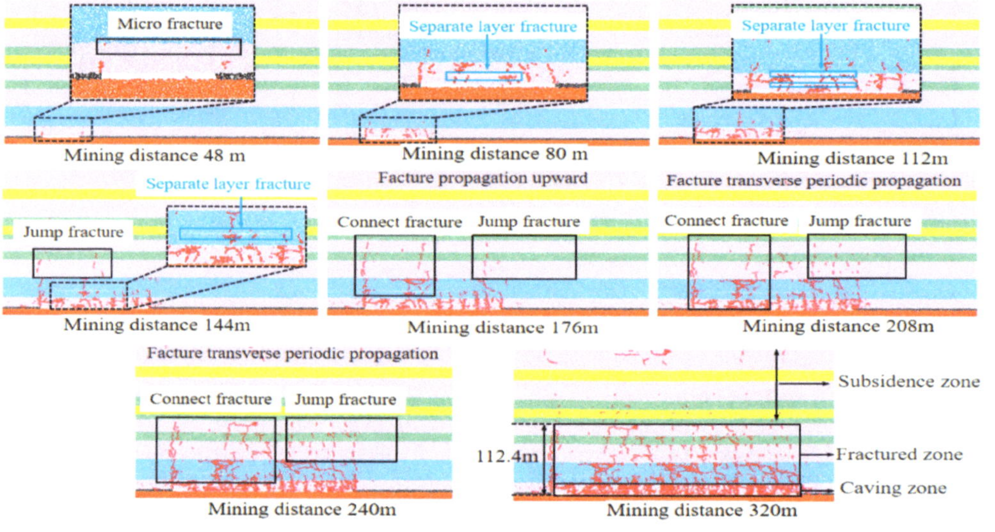

Figure 4. Damage development of coal mining face overburden rock with very different mining distances.

A statistical analysis was conducted on the development of overburden fractures and the height of water-conducting fracture zones during the advancement of the coal mining face, with results shown in Figure 5. It reveals that the number of fractures and the height of the water-conducting fracture zones gradually increase as the mining area progresses forward, and there is a good correlation between the growth rates of fractures and the height of the water-conducting zones. Based on these rates, the process can be divided into three stages: slow growth, rapid growth, and steady growth. From 0 m to 64 m of advancement, the slow growth stage is characterized by a small goaf area and low mining disturbance, with fractures primarily distributed in the immediate and main roof layers. From 64 m to 176 m, the rapid growth stage occurs, with multiple sets of vertically through-going fractures interconnecting laterally in the immediate and main roof layers, and vertical fractures in the higher overburden layers rapidly developing and gradually interconnecting, significantly increasing the distribution range and number of overburden fractures. From 176 m to 340 m, the steady growth stage features lateral periodic expansion of fractures, with the height of fracture development remaining essentially unchanged, thus transitioning from nonlinear to linear growth in the number of fractures.

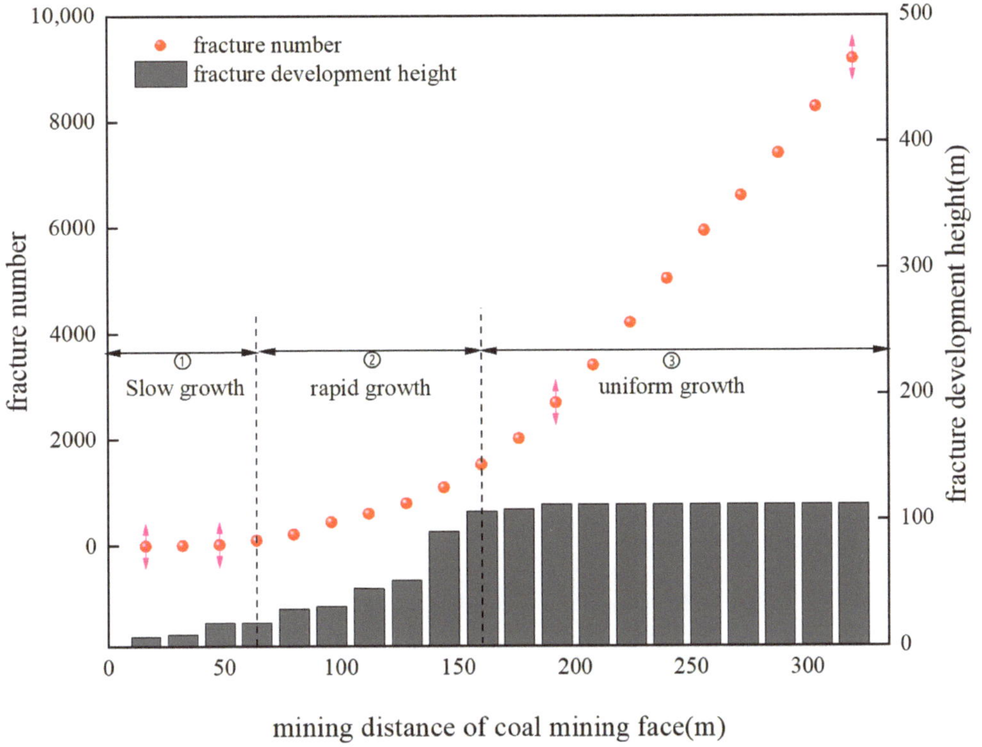

Figure 5. Changes in the number and development height of overburden fractures.

3.2.2. Evolution Characteristics of Overburden Rock Force Chain

Mining disturbances in the coal seam disrupt the equilibrium of the initial force chain system. Within the disturbed range, numerous force chains experience fracture and re-contact, forming a new network of force chains. The morphology and distribution characteristics of this network reflect the stress distribution in the overburden and can reveal the mechanisms of fracture formation and expansion through analysis of the overburden force chain network features. Figure 6 illustrates the evolution of the overburden force chain network at various distances of the coal mining face advancement. In the figure, black lines represent compression chains formed under particle compression, while yellow lines represent tension chains formed under particle tension; thicker lines indicate greater contact forces.

As shown in Figure 6, before coal mining, the force chain network was primarily oriented vertically under the influence of gravity. At 48 m into the coal mining face advancement, the force chain direction in a small overlying area of the goaf shifts, forming a horizontal bundle of strong force chains. Multiple rock layers together form a multi-layered strong force chain arch, with the feet of the arch on both sides of the goaf. The maximum contact force within the arch is 1616 KN. The overlying rock load is supported by this strong force chain arch and transferred to the coal layers in front of and behind the mined area. In areas further from the goaf, rock layer force chain deflection is less apparent, and chains primarily develop vertically. Inside the overburden, pressure chains are distributed in rock layers unaffected by mining disturbance and within the strong force arch, while tension chains are located in the middle to lower parts of the arch crown and the middle to upper sides of the arch feet.

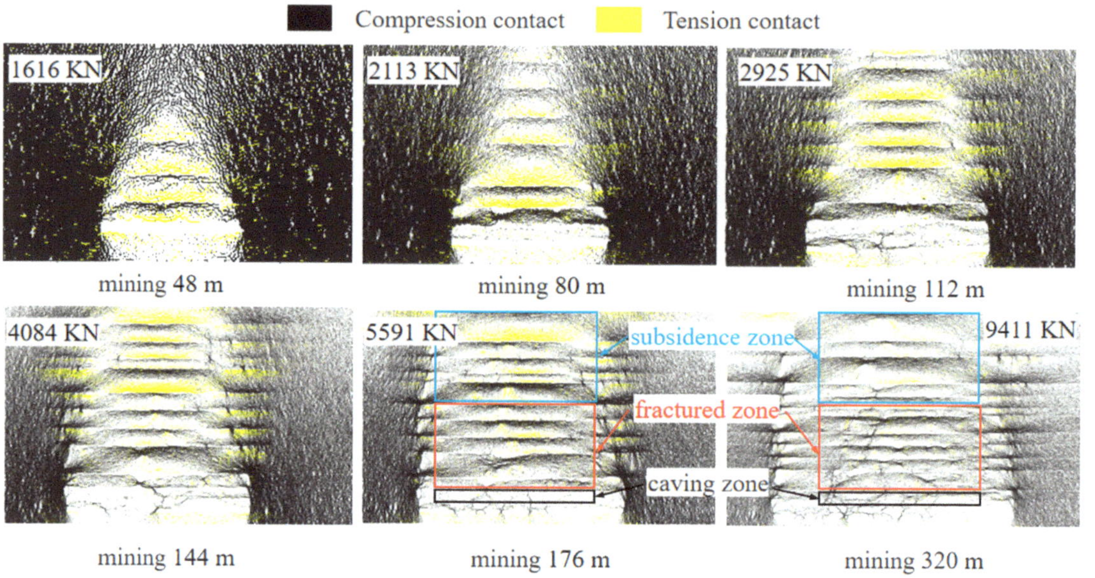

Figure 6. Evolution process of overburden force chain.

At an 80 m advancement of the coal mining face, the strong force arch in the lower rock layers exceeds its own load-bearing limit due to the overlying load, leading to rock layer fracture and the formation of new load-bearing force chains through hinge jointing. After the rock layers fracture, the tension chains disappear, and pressure concentration points emerge at the hinged locations of the rock blocks, with a maximum contact force of 2113 KN. A new strong force arch appears in the higher rock layers, indicating that the height of the force arches increases with the expanded mining range, while the strong force chains in the collapsed rock bodies within the goaf disappear, degrading into unbonded weak chains. As the coal mining face progresses to 112 m, 144 m, and 176 m, the strong force arches gradually deteriorate from the bottom up along the middle of the arch crown and the outer sides of the arch feet. At these points of destruction, particles make secondary contact to form new load-bearing structures, with maximum contact forces increasing to 2925 KN, 4084 KN, and 5591 KN, respectively. The strong force arches in the higher rock layers have developed to the top of the model, gradually expanding the mining impact area.

As the coal mining face advances, the vertical structure of the overburden force chain remains essentially unchanged, consistent with the distribution of overburden fractures, and expands laterally in a periodic manner with the advancement of the mining area. The distribution of the force chain network also reveals the characteristics of the three zones of overburden: in the caving zone, the overall strength of the force chains is weak, with strong vertical force chains formed by the compaction of broken rock blocks; the fracture zone contains numerous load-bearing horizontal force chain bundles, with a few vertical strong force chains in the far area behind the mining field, where the contact strength is significantly stronger than in the caving zone, and there are many points of stress concentration; in the bending and sinking zone, rock layer fractures are not interconnected, the integrity of the rock layers is not completely lost, and the morphology of the force chains is that of multiple strong arches, with many tensile chains distributed below the arch crown and above the arch feet. The contact strength gradually increases with the advancement distance, reaching a maximum contact force of 9411 KN at 320 m.

3.3. The Propagation Mechanism of Water-Conducting Fracture in Overburden Rock

As shown in Figure 7, during the coal seam mining process, tensile stress concentrations appear in the lower middle area and the upper side areas of the suspended roof. Due to the characteristic of rock being strong in compression but weak in tension, the force chains in these areas are the first to break, creating tensile microcracks. In these microcrack areas, the strong force chains disappear or weaken, forming a weak force chain zone. After the appearance of microcracks, part of the stress is released while the remaining stress is transferred to the surrounding force chains. Under the combined effects of stress transfer and accumulation, the force chain strength at the ends of the microcracks increases, forming strong force chain zones. When the stress concentration in the strong force chain zones exceeds the contact strength limit between particles, fractures occur. At this point, multiple microcracks connect and penetrate the entire rock layer, forming through-going fractures with water-conducting properties.

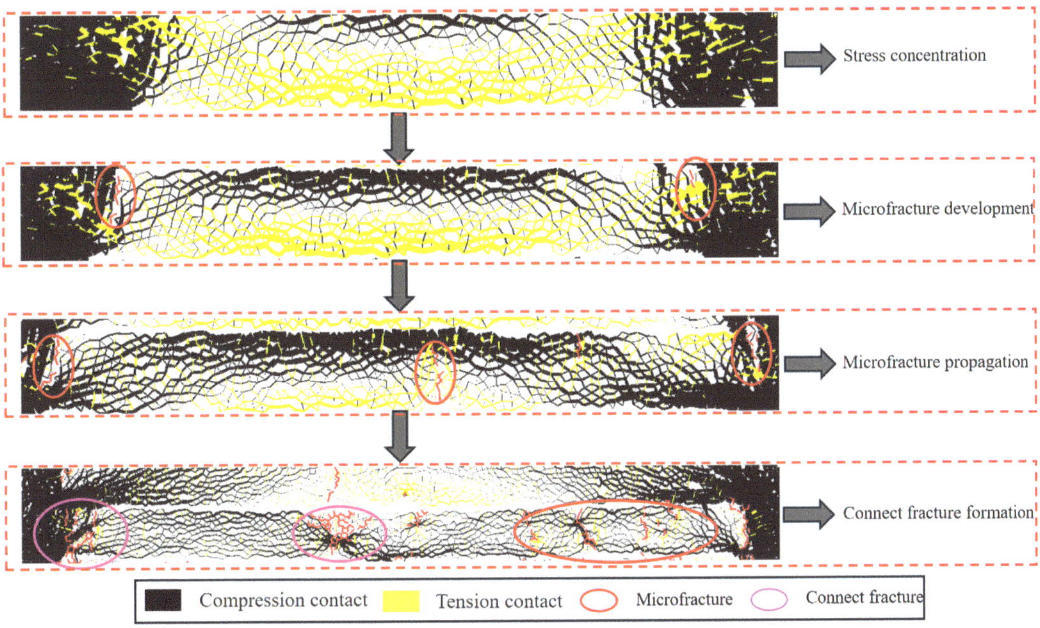

Figure 7. Crack propagation mechanism of overburden rock.

The overburden above the mined-out area is composed of multiple layers of different rock types. However, the fracture expansion patterns within the individual rock layers of different lithologies are consistent. Therefore, during the advancement of the mining area, unconnected microcracks, termed leapfrog fractures, coexist simultaneously in multiple groups of rock layers. These fractures typically develop in higher rock strata and within a small range behind the coal mining face. As the coal mining face progresses, these leapfrog fractures gradually expand and interconnect to form water-conducting fracture channels. The rupture of the overburden typically lags behind the coal mining face, with higher strata showing more pronounced lag, resulting in the formation of the irregular water-conducting fracture channels shown in Figure 8. Overall, the analysis reveals that during coal mining, the formation process of the overburden water-conducting fracture channel progresses through stages of "local microfractures, leapfrog fractures, interconnected fractures, to water-conducting fractures".

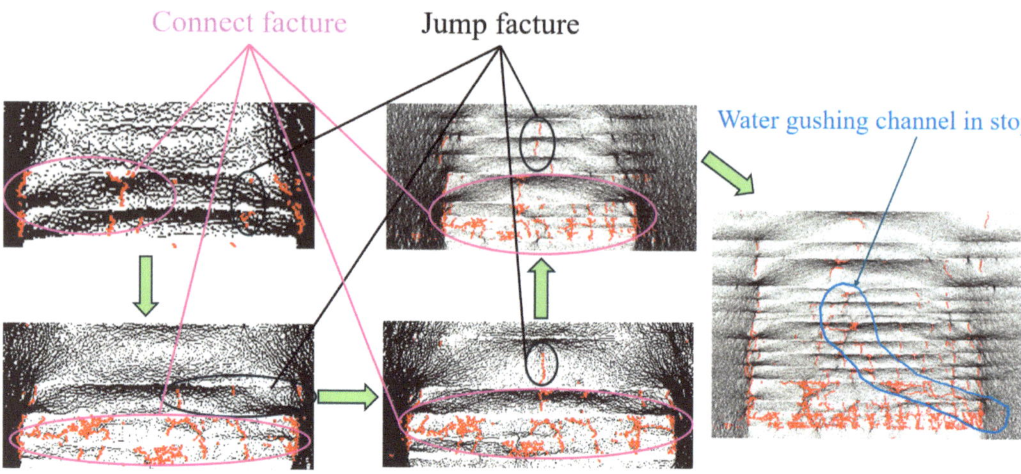

Figure 8. Formation process of water-conducting fracture channel in overburden rock.

4. Physical Simulation of Water-Conducting Fracture Development Height

This experiment used a two-dimensional physical similarity simulation to model the destruction of overlying strata during the mining process of a moderately deep coal seam, calculating the development height of the "two zones" and summarizing their developmental patterns. Based on the actual mining conditions of the 2102 coal mining face in the 21 mining area of Balasu Coal Mine, a two-dimensional experimental model was constructed using similar materials to simulate the deformation patterns of the overlying rock mass affected by coal mining disturbances. The study analyzed the initial pressure step, periodic pressure step, and pressure distribution under the specific geological and mining conditions. Observations were made on the deformation patterns of the overburden as the coal seam advanced, analyzing the impact of base rock layer rupture instability and complete deformation of the overburden. The study also examined the relative relationships of development heights for different mining heights, stope ratios, and fracture ratios to determine the development heights of the "two zones".

4.1. Physical Model Construction and Scheme Design

The experimental design was based on the borehole stratigraphic charts and physical and mechanical parameters of the rocks surrounding the coal seam at the 2102 coal mining face in the 21 mining area of the Balasu Coal Mine, as shown in Figure 9. Given constraints due to rock layer stability and model height, a frame measuring 2.80 m in length, 0.20 m in width, and 1.15 m in height was used. The experiment employed a geometric similarity constant of 1:200 to simulate the development of the "two zones" in a 220 m overburden layer of the 3.2 m high coal seam buried at 483.64 m. A three-dimensional speckle system was utilized for comprehensive monitoring of the model. Additionally, 36 observation lines were set in the vertical direction and 117 in the horizontal direction, totaling 4212 monitoring points to form a comprehensive observation network for the two-dimensional experimental model.

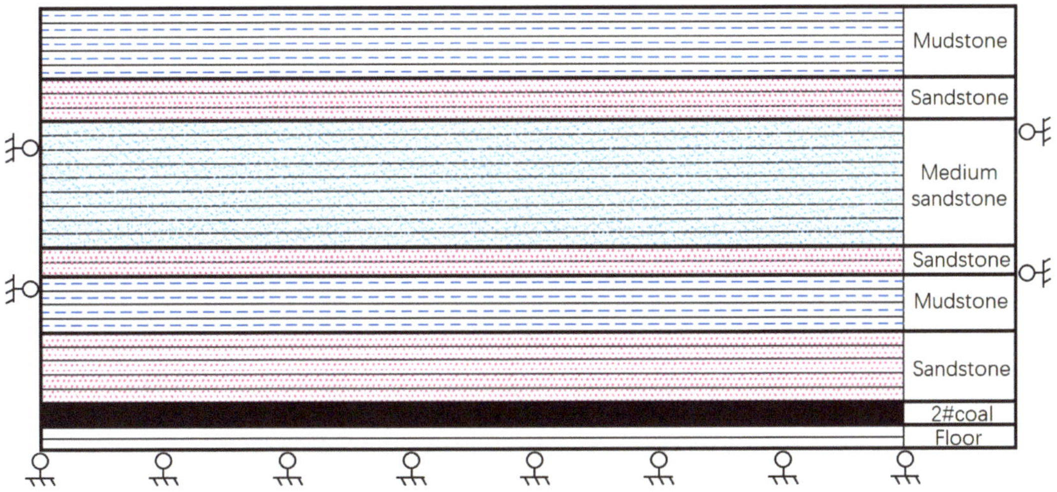

Figure 9. Physical simulation platform and model construction.

4.2. Overburden Rock Migration Characteristics and Development Height of Two Zones

The migration pattern of overlying strata at different advancing distances of the working face is shown in Figure 10. When the coal mining face advanced to 40 m, the coal seam roof first experienced collapse. At 50 m, a distinct caving zone emerged in the overlying rock layer of the coal seam roof, and the caving area developed forward and towards the surface along the mining direction. By 70 m, compared to the state at 50 m, the vertical development speed of the caving area slowed down. Influenced by mining disturbances and periodic pressure arches, there was significant horizontal expansion in the caving area along the mining direction of the coal seam. Additionally, subsidence deformation in the overlying rock layer of the caving area and visibly developing cracks indicated that a water-conducting fracture zone had formed and continued to develop with mining. At 140 m, an initial pressure-induced "saddle-shaped" caving area formed under the coal seam roof, with the caving zone reaching its maximum height and displaying a symmetrical distribution characteristic of an initial pressure arch. Additionally, the caving zone continued to develop toward the surface, forming an elastic expansion area. Within this area, the rock layers did not fracture, but clear stratification was evident in both the caving zone and the water-conducting fracture zone, indicating the distinctive development characteristics of the "two zones." The vertical development height of the caving zone reached its maximum, with virtually no further vertical growth, while the water-conducting fracture zone expanded with coal seam mining. As the coal mining face advanced to 180 m, the initial pressure arch in the coal seam roof extended maximally in the mining direction, with the caving zone also reaching its maximum development height, particularly near the cutting face, marking the initial pressure arch. Additionally, horizontal fractures developed above the caving zone, extending the water-conducting fracture zone towards the surface and creating a distinct elastic expansion area in the overburden of the mined coal seam. At a 200 m advancement of the coal mining face, the caving zone and the water-conducting fracture zone exhibited pronounced stratification, indicating the developed characteristics of the "two zones". The vertical development height of the caving zone reached its maximum, with virtually no vertical growth, while the water-conducting fracture zone predominantly expanded vertically. At this stage, the horizontal fractures (delamination zone) were primarily located above the caving zone and below the water-conducting fracture zone. When the coal mining face advanced to 280 m, the maximum vertical deformation at the top of the caving zone in the overlying rocks was 20 mm, with the water-conducting fracture zone almost reaching its maximum development height approximately 100 m from the coal seam roof, where the maximum vertical deformation within the water-conducting fracture zone was 14 mm. When the coal mining face advanced to 310 m, periodic pressure arches formed in the overlying strata of the coal seam roof, with a periodic pressure step distance of about 40 m. At this point, both the caving zone and the water-conducting fracture zone developed to their maximum heights, with the maximum height of the water-conducting fracture zone being 100 m from the coal seam roof, located near the cutting eye. The upper part of the water-conducting fracture zone in the mined overlying rock showed bending deformation, but no visible vertical or horizontal mining fractures were observed, indicating that the upper strata of the water-conducting fracture zone underwent integral bending deformation and collapse. When the entire simulation of the coal mining face reached 480 m, the water-conducting fracture zone developed laterally to its largest area, with the stable maximum height of the water-conducting fracture zone being about 100 m during the periodic destruction process of the mined overlying rock of the coal seam roof. The coal seam roof exhibited significant periodic pressure-induced collapse. The deformation within the mined overlying rock, particularly the subsidence towards the surface, shows a significant deceleration. The deformation of the mined overlying rock above the water-conducting fracture zone is predominantly characterized by bending and subsidence, without any noticeable mining-induced fractures.

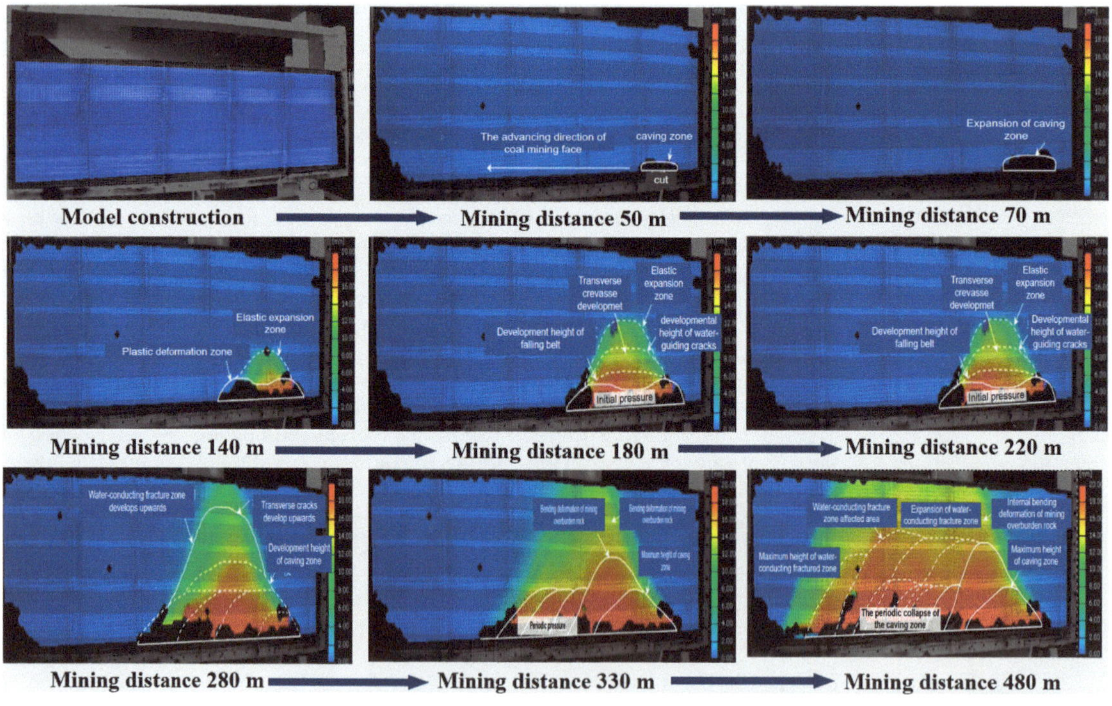

Figure 10. Failure characteristics of overburden rock with different advancing distances.

Based on the similar material model experiment, with a coal seam mining height of 3.3 m and the coal mining face fully mined, the maximum development height of the "two zones" was determined to be approximately 100 m, with the periodic pressure mining step distance ranging from 30 m to 40 m. The critical threshold between fully mined and insufficiently mined conditions at the coal mining face is located within a coal seam advancement range of 280 m to 330 m. In the early stages of the water-conducting fracture zone formation, the development speed of the caving zone along the direction of the coal mining face's advancement was significantly higher than its development speed toward the surface. This meant that the vertical development of the caving zone slowed down. The caving zone, influenced by mining disturbances and periodic pressure arches, exhibited significant horizontal expansion along the mining direction of the coal seam. Subsidence deformation and lateral fractures appeared in the overlying rock layers of the caving zone. As mining advanced, the height of the caving zone in the coal seam roof reached its maximum, forming an initial pressure-induced "saddle-shaped" caving area, with the maximum height of the caving zone displaying a symmetrical distribution, characteristic of the initial pressure arch. When distinct stratification appears in the caving zone and the water-conducting fracture zone, fully revealing the development characteristics of the "two zones", the vertical development height of the caving zone reaches its maximum, and the vertical development speed approaches zero. As mining progresses, expansion of the water-conducting fracture zone occurs along with the coal seam being mined. Simultaneously, the initial pressure arch in the coal seam roof extends maximally in the direction of mining, and the height of the caving zone develops to its maximum, with the highest point located near the cutting eye, marking the initial pressure arch. Additionally, horizontal fractures develop above the caving area, the water-conducting fracture zone extends towards the surface, and a noticeable elastic expansion area appears in the overlying strata of the mined coal seam.

5. Field Measurement of Water-Conducting Fracture Development Height

5.1. Detection Scheme Design

Based on the geological conditions and the current mining status of the 2102 coal mining face, the surface subsidence basin within 600 m of the cutting eye has stabilized, providing suitable conditions for conducting exploratory measurements of the "water-conducting fracture zone". Consequently, the exploration site for the fracture zone was chosen at a location 500 m from the cutting eye. Three different methods were selected for investigating the development height of the fracture zone: flush fluid consumption monitoring, borehole television imaging, and hydraulic testing. The borehole locations are shown in Figure 11, with two boreholes, LT-2 and LT-3, constructed above the 2102 goaf, at depths of 478 m and 475 m, respectively, serving as observation wells for the fracture zone. Borehole LT-1, located in the unmined area north of the 2102 coal mining face as a pre-mining comparison borehole, reaches a depth of 480 m. The hydraulic testing equipment used was the "Two Zone Height Tester", independently developed by the Xi'an Research Institute, and the borehole television inspection used a SYKJ 17 model viewer.

5.2. Detection Scheme Design Detection Result Analysis

5.2.1. Flushing Fluid Leakage Observation

The results of the flush fluid loss and water level observations for each borehole are shown in Figures 12 and 13. From the figures, it can be seen that:

(1) In borehole LT-1, the flush fluid loss per unit time in the bedrock section varied from 0.0080 to 0.093 L/s·m, with an average value of 0.036 L/s·m. Throughout the observation period, the flush fluid loss did not increase with the depth of the borehole, and the fluid circulation remained normal without any interruption or loss. The water level in this borehole gradually decreased with increasing depth, ranging from 0.42 to 23.98 m. The water level changes were normal throughout the entire borehole, with no sudden drops or instances of the borehole running dry.

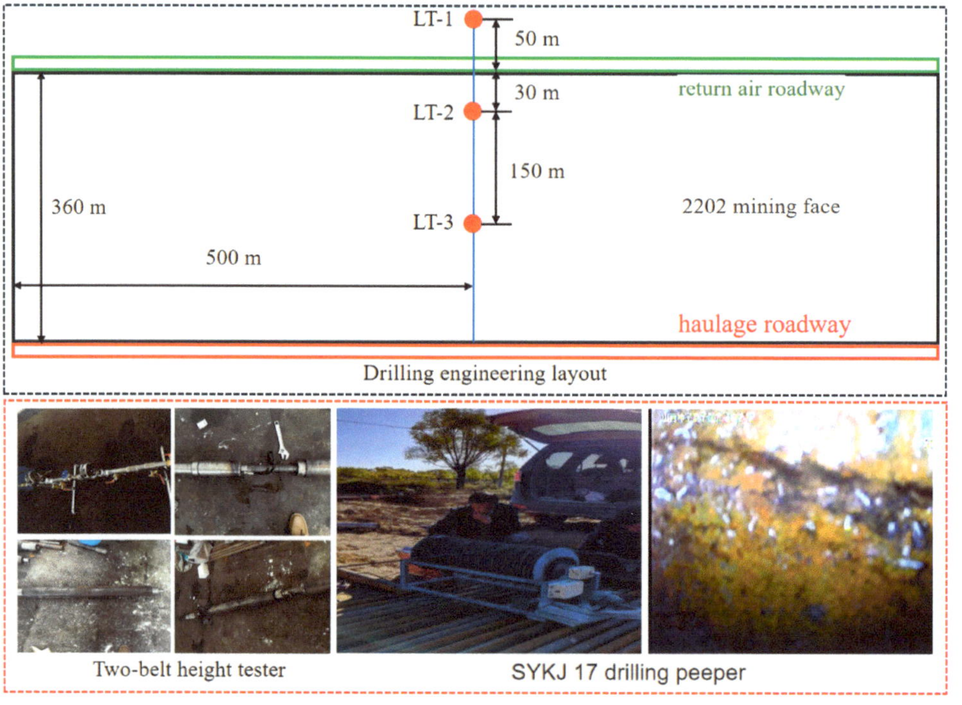

Figure 11. Exploration engineering and equipment for water flowing fractured zone in overlying strata.

(2) In borehole LT-2, the flush fluid loss in the bedrock section varied from 0.0088 to 1.73 L/s·m, with an average value of 0.28 L/s·m. In the sand and soil layers, the fluid loss did not vary significantly. At a depth of 373.40 m, the consumption began to increase, and by 402 m, all the flush fluid was lost, with no further circulation. The water level in this borehole gradually decreased with increasing depth, ranging from 1.90 to 31.24 m. The water level remained relatively stable in the upper bedrock section, but after drilling through the 368.99–373.40 m section, the water level dropped from 31.24 m to 156.78 m, and the consumption rate increased from 0.071 L/s·m to 0.91 L/s·m. By the time drilling reached 402 m, all flush fluid was lost. Based on these characteristics, the top boundary of the water-conducting fracture zone in borehole LT-2 is determined to be at a depth of 373.40 m.

(3) In borehole LT-3, the flush fluid loss in the bedrock section varied from 0.000593 to 2.10 L/s·m, with an average value of 0.297 L/s·m. At a depth of 365.10 m in the bedrock section, the consumption began to increase, and by 398 m, all the flush fluid was lost. The water level in this borehole gradually decreased with increasing depth, ranging from 1.13 to 39.40 m in the bedrock section. The water level remained relatively stable in the upper bedrock section, but after drilling through the 360.42–365.10 m section, the water level dropped from 39.40 m to 253.64 m, and the consumption rate increased from 0.048 L/s·m to 1.139 L/s·m. By the time drilling reached 398 m, all flush fluid was lost. Based on these characteristics, the top boundary of the water-conducting fracture zone in borehole LT-3 is determined to be at a depth of 365.10 m.

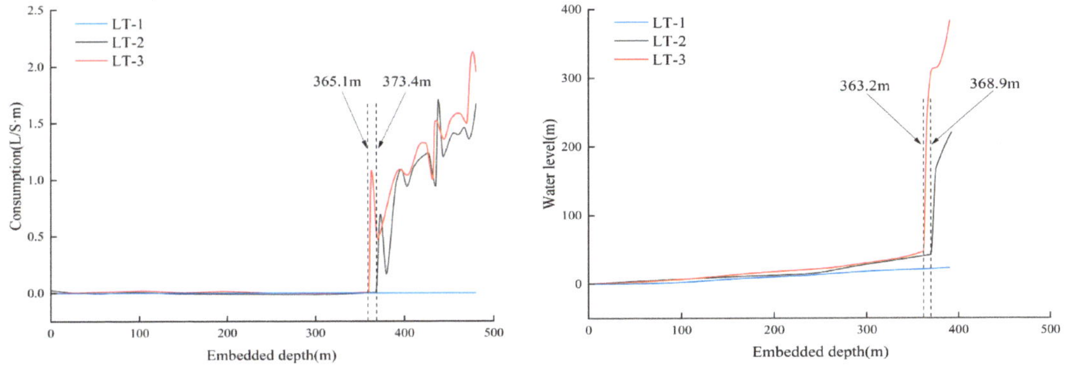

Figure 12. Flushing fluid leakage and water level varies with buried depths.

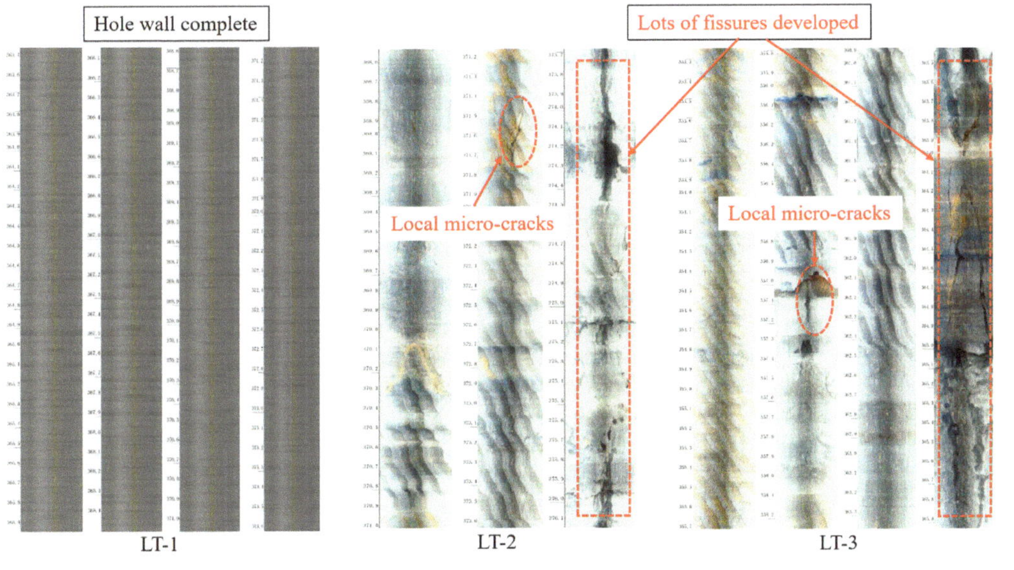

Figure 13. The development of cracks in the hole.

5.2.2. In-Hole TV Imaging Observation

Borehole inspection requires the borehole to remain dry or the borehole wall to be free of water flow. Therefore, during this drilling operation, water-stopping measures were implemented in the weathered bedrock and above. After the borehole completely lost fluid and became dry, borehole television inspection was conducted, as shown in Figure 13. In borehole LT-1, the borehole wall was generally intact, with no significant fractures observed. In borehole LT-2, local fractures were observed at a depth of 371.40 m, with small fracture width and distribution range. Significant destructive fractures began to develop below 373.70 m, mainly vertical fractures with some horizontal fractures. In borehole LT-3, vertical fractures appeared at a depth of 357.10 m, with a small fracture width and distribution range. Significant destructive fractures began to develop below 363.40 m, mainly vertical fractures. Compared to borehole LT^{-2}, the fractures in borehole LT-3 were more developed and extensive, indicating more intense movement and higher degrees of overlying rock damage in the central goaf.

5.2.3. Pump-In Test

Based on the observations of flush fluid loss and borehole television results, there are no significant mining-induced fractures above a depth of 357.10 m. Therefore, the section from 330.00 m to the end of the borehole was selected for hydraulic testing, with a test conducted for every meter drilled. The permeability of the rock mass is typically indicated by the permeability coefficient (k) and the permeability rate (q). The variation in k and q values with depth for each borehole is shown in Figure 14.

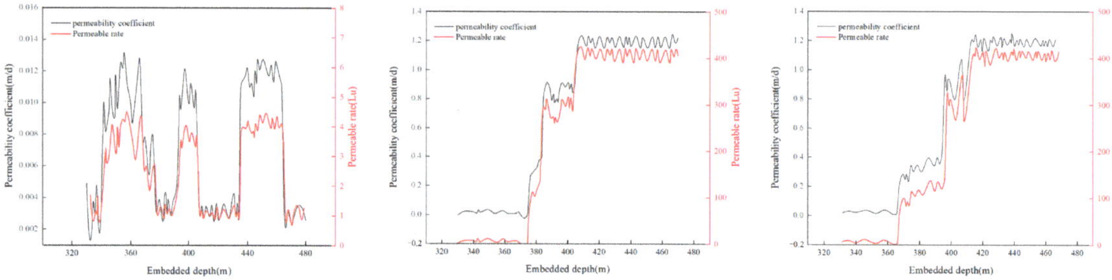

(**a**) Hole LT-1 (**b**) Hole LT-2 (**c**) Hole LT-3

Figure 14. Results of water pressure test.

The hydraulic test section of borehole LT-1 ranges from 330.00 m to 480.00 m, comprising a total of 150 test sections. As shown in Figure 14a, the permeability coefficient of the bedrock section in borehole LT-1 ranges from 0.0024 to 0.0134 m/d, with an average of 0.0073 m/d. The permeability rate ranges from 0.83 to 4.61 Lu, with an average of 2.51 Lu. Based on the classification of rock and soil permeability, the original rock mass permeability is determined to range from slightly permeable to weakly permeable.

The hydraulic test section of borehole LT-2 ranges from 330.00 m to 470.00 m, comprising a total of 140 test sections. As shown in Figure 14b, the k and q values of borehole LT-2 increase non-linearly with depth. The k and q value curves stabilize above a depth of 374 m, increasing to 0.97 m/d and 287.63 m/d at 374 m, and further to 1.33 m/d and 379.42 m/d at 408 m, before gradually stabilizing. The permeability grade of the overlying strata changes from weakly permeable to strongly permeable at a depth of 374 m.

The hydraulic test section of borehole LT-3 ranges from 330.00 m to 468.00 m, comprising a total of 138 test sections. As shown in Figure 14c, the k and q value curves for borehole LT-3 exhibit a similar non-linear growth trend to those of borehole LT-2. The k and q value curves stabilize above a depth of 366 m, increasing to 1.06 m/d and 308.26 m/d at 366 m, and further to 1.38 m/d and 386.34 m/d at 416 m, before gradually stabilizing. The permeability grade of the overlying strata changes from weakly permeable to strongly permeable at a depth of 366 m.

5.2.4. Comprehensive Determination of Height of Water Flowing Fractured Zone

The development heights of water-conducting fractures obtained by different observational methods and their comparisons are shown in Tables 2 and 3. According to Table 2, the three methods provide closely aligning results for the height of the water-conducting fracture zone at the 2102 working face, ranging from 103.68 to 107.58 with a fracture ratio of 31.42 to 32.60. Table 3 presents the fracture zone heights obtained by different research methods, revealing that the maximum discrepancies between numerical simulations, physical simulations, and actual measurements are 4.2% and 7%, respectively. The field measurements confirm the validity and accuracy of the theoretical research described above.

Table 2. Detection results of development height of water flowing fractured zone.

Drilling	Elevation of Bore Hole/m	Buried Depth of Floor/m	Floor Level/m	Critical Depth of Water -Conducting Fracture Zone/m	Height of Fractured Water-Conducting Zone/m	Height Mining/m	Crack Production Ratio
LT-2	+1185.3833	477.08	+705.00	373.40	103.68	3.30	31.42
LT-3	+1180.9764	472.68	+705.00	365.10	107.58	3.30	32.60

Table 3. Comparison of water-conducting fracture development height results.

Research Method	Height of Fractured Water-Conducting Zone/m	Crack Production Ratio	Error/%
Numerical simulation	108	31.42	4.2%
Physical simulation	100	32.60	7%
Field test	103.68~107.58	103.68~107.58	/

6. Conclusions

(1) Within the range of mining influence, the direction of force chains in the rock strata undergoes horizontal deflection, with multiple layers forming multiple strong chain arches. Tensile stress concentration areas between layers first develop tensile cracks. Under the effect of stress transmission and superposition, the local stress concentration at the ends of tensile micro-cracks increases, leading to the development of jump fractures and through fractures. The formation process of water-conducting fractures is "local micro-fractures–jump fractures–through fractures–water-conducting fractures" with the maximum height of the water-conducting fracture zone being approximately 108 m.

(2) By observing the water loss in three underground boreholes, conducting borehole television observations, and combining the results of hydraulic tests, the height of the water-conducting fracture zone was comprehensively determined to be 103.68–107.58 m, with a fracture-to-mining ratio of 31.42–32.60. These results are not significantly different from those obtained through numerical and physical simulations. The findings of this study can provide theoretical guidance and scientific basis for coal mine water hazard prevention and control under similar geological conditions.

(3) By conducting water injection loss measurements, borehole television observations, and hydrostatic testing on three underground boreholes, the height of the water-conducting fracture zone was determined to range between 103.68 and 107.58, with a fracture ratio of 31.42 to 32.60, closely matching the numerical and physical simulation results. This research provides theoretical guidance and a scientific basis for the prevention of water hazards in coal mines under similar geological conditions.

(4) The numerical and physical models developed in this paper simplify the actual geological conditions on-site. Future research can focus on developing more accurate simulation technologies to better reflect real situations under complex geological conditions.

Author Contributions: Conceptualization, D.W.; Software, H.G.; Investigation, C.W.; Resources, H.G.; Data curation, D.W. and C.W.; Writing—original draft, H.W., H.Z. and Y.G. All authors have read and agreed to the published version of the manuscript.

Funding: This research received no external funding.

Data Availability Statement: The original contributions presented in the study are included in the article, further inquiries can be directed to the corresponding author.

Conflicts of Interest: Authors Wei Dong and Chungang Wang were employed by Shanxi Yanchang Petroleum and Mining Co., Ltd. The remaining authors declare that the research was conducted in the absence of any commercial or financial relationships that could be construed as a potential conflict of interest.

References

1. Li, W.; Qiao, W.; Li, X.; Sun, R. Chacteristics of water disaster, evaluation methods and exploration direction for control-ling ground water in deep mining. *J. China Coal Soc.* **2019**, *44*, 2437–2448.
2. Wang, S. Reflections on the energy status of coal bodies and green mining in China. *J. China Coal* **2020**, *46–48*, 11–16.
3. Wu, Q. Progress, problems and prospects of prevention and control technology of mine water and reutilization in China. *J. China Coal Soc.* **2014**, *39*, 795–805.
4. Tan, Y.; Guo, W.; Yang, D.; Bai, E. Analysis on height of "two zones" under subcritical mining in shallow coal seam with hard roof. *J. Min. Saf. Eng.* **2017**, *34*, 845–851.
5. Helong, G.; Xingping, L.; Ming, T.; Aliakbar, M.; Qunlei, Z. Dynamic mechanical mechanism and optimization approach of roadway surrounding coal water infusion for dynamic disaster prevention. *Measurement* **2023**, *223*, 113639. [CrossRef]
6. Gao, B.; Liu, Y.; Pang, J.; Yuan, T. Detection and analysis of height of water flowing fractured zone in under water mining. *Chin. J. Rock Mech. Eng.* **2014**, *33*, 3384–3390.
7. Chen, Y.; Zhu, S. Determination of caved and water-conducting fractured zones of "two soft and one hard" unstable coal seam. *Acta Geod. Geophys.* **2020**, *55*, 451–475. [CrossRef]
8. Hebblewhite, B. Fracturing, caving propagation and influence of mining on groundwater above longwall panels—A review of predictive models. *Int. J. Min. Sci. Technol.* **2020**, *30* (Suppl. S1), 49–54. [CrossRef]
9. Ye, Q.; Wang, G.; Jia, Z.Z.; Zheng, C.; Wang, W. Similarity simulation of mining-crack-evolution characteristics of overburden strata in deep coal mining with large dip. *J. Pet. Sci. Eng.* **2018**, *165*, 477–487. [CrossRef]
10. Wang, G.; Wu, M.; Wang, R.; Xu, H.; Song, X. Height of the mining-induced fractured zone above a coal face. *Eng. Geol.* **2017**, *216*, 140–152. [CrossRef]
11. Xu, Y.; Ma, L.; Khan, N.M. Prediction and maintenance of water resources carrying capacity in mining area—A case study in the Yu-Shen mining area. *Sustainability* **2020**, *12*, 7782. [CrossRef]
12. Liu, S.; Li, W. Zoning and management of phreatic water resource conservation impacted by underground coal mining: A cause study in arid and semiarid areas. *J. Clean. Prod.* **2019**, *224*, 677–685. [CrossRef]
13. Singh, M.M.; Kendorski, F.S.T. The effects of longwall coal mining on overlying aquifers. In *Mine Water Hydrogeology and Geochemistry*; Geological Society Special Publication; Younger, P.L., Robins, N.S., Eds.; Geological Society of London: London, UK, 2022; Volume 198, pp. 17–45.
14. Coe, C.J.; Stowe, S.M. Strata-movement concepts and the hydrogeological impact of underground coal mining. *Ground Water* **1986**, *24*, 507–515.
15. Boothc, J. A reappraisal of investigations into strata permeability changes associated with longwall mining. *Mine Water Environ.* **1983**, *2*, 1–14.
16. Jia, C.; Lai, X.; Cui, F.; Xu, H.; Zhang, S.; Li, Y.; Zong, C.; Luo, Z. Mining Pressure Distribution Law and Disaster Prevention of Isolated Island Working Face Under the Condition of Hard "Umbrella Arch". *Rock Mech. Rock Eng.* **2024**, 1–19. [CrossRef]
17. Zhang, L.; Lai, X.; Pan, J.; Shan, P.; Zhang, Y.; Zhang, Y.; Xu, H.; Cai, M.; Xi, X. Experimental investigation on the mixture optimization and failure mechanism of cemented backfill with coal gangue and fly ash. *Powder Technol.* **2024**, *440*, 119751. [CrossRef]
18. Huang, B.; Xie, P.; Wu, Y.; Lin, W.; Luo, S.; Wang, S.; Wen, Z.; Chen, J. The efect of overlying rock fracture and stress path evolution in steeply dipping and large mining height stope. *Géoméch. Geophys. Geo-Energy Geo-Resour.* **2024**, *10*, 1–19. [CrossRef]
19. Wei, J.; Wang, S.; Song, S.; Sun, Q. Numerical simulation on evolution law of overburden fractures and surface cracks in crossing ditch miming of shallow coal seam. *Coal Geol. Explor.* **2022**, *50*, 67–75.
20. Wang, C.; Zhang, C.; Zhao, X.; Liao, L.; Zhang, S. Dynamic structural evolution of overlying strata during shallow coal seam longwall mining. *Int. J. Rock Mech. Min. Sci.* **2018**, *103*, 20–32. [CrossRef]
21. Li, X.F.; Li, X.; Li, H.B.; Zhao, J. Dynamic tensile behaviours of heterogeneous rocks: The grain scale fracturing characteristics on strength and fragmentation. *Int. J. Impact Eng.* **2018**, *118*, 98–118. [CrossRef]
22. Li, X.F.; Zhang, Q.B.; Li, H.B.; Zhao, J. Grain-based discrete element method(gb-dem) modelling of multi-scale fracturing in rocks under dynamic loading. *Rock Mech. Rock Eng.* **2018**, *51*, 3785–3817. [CrossRef]
23. Dai, J.; Liu, J.; Liu, C.; Yang, J.; Xue, F.; Tang, Y.; Liu, J.; Liu, D.; Dai, S. Regional Characteristics of Seismicity Associated with Hydraulic Fracturing in the Southern Sichuan Basin of China. *Rock Mech. Rock Eng.* **2024**, 1–26. [CrossRef]
24. Dai, J.; Liu, J.; Zhou, L.; He, X. Crack Pattern Recognition Based on Acoustic Emission Waveform Features. *Rock Mech. Rock Eng.* **2023**, *56*, 1063–1076. [CrossRef]
25. Zhang, T.; Yu, L.; Xu, F.; Ju, M.; Pu, H.; Li, W.; Liu, J. Influence of Mesoscopic Defects on the Mechanical Behaviour of Granites Based on a Three-Dimensional Multilevel Force Chain Network. *Rock Mech. Rock Eng.* **2024**, 1–21. [CrossRef]
26. Chen, P.; Kong, Y.; Yu, H. Research on the Calibration Method of Microparameters of a Uniaxial Compression PFC2D Model for Rock. *Chin. J. Undergr. Space Eng.* **2018**, *14*, 1240–1249.
27. Zhang, T.; Yu, L.Y.; Su, H.J.; Zhang, Q.; Chai, S. Experimental and numerical investigations on the tensile mechanical behavior of marbles containing dynamic damage. *Int. J. Min. Sci. Technol.* **2022**, *32*, 89–102. [CrossRef]

28. Yang, S.; Tian, W.; Huang, Y.; Ranjith, P.G.; Ju, Y. An Experimental and numerical study on cracking behavior of brittle sandstone containing two non-coplanar fissures under uniaxial compression. *Rock Mech. Rock Eng.* **2016**, *49*, 1497–1515. [CrossRef]
29. Xu, Y.; Dai, F.; Xu, N.W.; Zhao, T. Numerical investigation of dynamic rock fracture toughness determination using a semi-circular bend specimen in split hopkinson pressure bar testing. *Rock Mech. Rock Eng.* **2016**, *49*, 731–745. [CrossRef]

water

MDPI

Article

Characteristics of Deformation and Damage and Acoustic Properties of Sandstone in Circular Tunnel Morphology under Varying Inundation Depths

Gang Liu [1,2,*], Shengxuan Wang [1], Dongwei Wang [1], Zhitao Yang [2] and Yonglong Zan [1]

1 Heilongjiang Ground Pressure and Gas Control in Deep Mining Key Laboratory, Heilongjiang University of Science and Technology, Harbin 150022, China; w13140438798@163.com (S.W.); 18297919986@163.com (D.W.); 17353693207@163.com (Y.Z.)

2 Baotailong New Material Co., Ltd., Qitaihe 154604, China; w120559779@163.com

* Correspondence: liugang@usth.edu.cn; Tel.: +86-189-4463-0110

Abstract: When water damage occurs in a mine, variations in the immersion levels of tunnels at different burial depths can be observed. There is a significant relationship between the stability of the surrounding rock and the depth of immersion. Therefore, studying the deformation and damage characteristics of sandstone with circular holes at varying immersion depths, along with their acoustic properties, plays a crucial role in maintaining the stability of water-rich roadways. The TAW-2000 press and static strain system were utilized to investigate the mechanical properties, crack evolution, and deformation field distribution of sandstone with circular holes at varying immersion depths. Additionally, this study analyzed the impact of immersion depth on the characteristic parameters of acoustic emission. The results indicate that immersion depth is negatively correlated with the compressive strength and modulus of elasticity of sandstone; as immersion depth increases, the duration of the compression and yield phases of the rock samples also increases, while the duration of the elastic phase remains relatively unaffected. Furthermore, greater immersion depths correspond to a decrease in the total number of cracks, although the proportion of tensile cracks increases, making the formation of secondary cracks less likely. The frequency of acoustic emission events (transient elastic waves generated by the formation, extension, or closure of tiny cracks within the rock) shows a closely correlated dynamic with the stress–time curve of the rock sample. The acoustic emission ringing counts generated by rock samples under submerged water conditions tend to stabilize with a slight increase before signs of rupture appear. Additionally, the cumulative total energy of acoustic emissions from the rock samples decreases as the water level rises. These research findings provide significant reference value for addressing issues related to water immersion and the extent of water saturation in roadways within rock engineering.

Keywords: immersion depth; porous sandstone; acoustic emission; deformation damage

Citation: Liu, G.; Wang, S.; Wang, D.; Yang, Z.; Zan, Y. Characteristics of Deformation and Damage and Acoustic Properties of Sandstone in Circular Tunnel Morphology under Varying Inundation Depths. *Water* **2024**, *16*, 2938. https://doi.org/10.3390/w16202938

Academic Editor: Fernando António Leal Pacheco

Received: 3 September 2024
Revised: 12 October 2024
Accepted: 14 October 2024
Published: 15 October 2024

1. Introduction

With the gradual development of mining activities in large water mines characterized by extremely complex hydrogeological conditions, the rock formations in these mines are typically situated in environments rich in groundwater. The presence of water significantly weakens the mechanical properties and structural stability of the rock [1,2]. When the groundwater level around the roadway rises or during heavy rainfall, water can easily penetrate and fill the roadway, leading to prolonged soaking and even saturation of the surrounding rock. Such environmental conditions severely impact the physical and mechanical properties of the rock, including its strength, bearing capacity, and deformation characteristics, thereby posing significant challenges to the stability of the roadway. Due to the varying water levels of the submerged roadway, the extent of damage to the roadway also differs significantly. To ensure that the sandstone in circular roadway formations maintains stability and safety under varying depths of submergence, this paper conducts an analytical study on the deformation, damage characteristics, and acoustic emission properties of sandstone containing circular roadways at different submergence depths. This research is highly significant for addressing issues related to the submergence of roadway peripheral rock in geotechnical engineering.

The issue of tunnel inundation is a significant focus in contemporary geotechnical engineering research and has garnered the attention of numerous scholars both domestically and internationally. Zhao Zenghui et al. [3] analyzed the stability of weakly cemented soft rock tunnel perimeter rock under the combined effects of a water environment and non-uniform ground pressure. They found that various factors exacerbate the instability of perimeter rock in high-humidity conditions. Li Bo et al. [4] investigated and managed the instability of coal bed roadways in complex water environments, concluding that the surrounding rock experiences a substantial reduction in strength due to water exposure. Additionally, some researchers examine the mechanical and acoustic changes in rock samples resulting from water body disturbances through experiments related to rock mechanics. Wu Baoyang et al. [5] concluded that the uniaxial compressive strength and modulus of elasticity of intact and fissured sandstone decrease with an increasing number of water immersion cycles, as demonstrated through uniaxial compressive experiments on sandstone subjected to multiple water immersions. Xia Dong et al. [6–8] investigated the mechanical characteristics and acoustic properties of water-saturated sandstone with varying immersion durations. Their findings, derived from uniaxial compression tests and acoustic emission tests, indicated that the pattern of cumulative acoustic emissions aligns with the internal damage evolution in the rock samples. Xiao Fukun et al. [9] examined the mechanical properties of dry and water-saturated pore rock samples under cyclic loading, concluding that the failure of rock samples in dry conditions is significantly less than that in water-saturated conditions. Tan Tao et al. [10] analyzed the mechanical properties of sandstone in both dry and water-saturated states using triaxial compressive tests, finding that damage in water-saturated sandstone is more pronounced under varying peripheral pressures, resulting in the formation of more cracks. Jiang Lisong et al. [11] studied the acoustic emission characteristics of shale with different water contents during uniaxial compression tests, concluding that shale specimens with varying water content generate acoustic emission signals during compression. Wang Yunfei et al. [12] determined that water-saturated sandstone can mitigate the effect of loading rate on the development of microscopic tensile fractures by investigating the influence of loading rate and water saturation on the mechanical behavior and microscopic damage characteristics of sandstone. Through experimental studies on the tensile properties of coal rock under pressurized water immersion, Yin Dawei et al. [13] concluded that the interaction between water and coal rock intensifies with increased water immersion pressure, compromising the initial energy storage structure of coal samples. Fang Jie et al. [14] examined the impact of water content on the strength damage and acoustic emission characteristics of muddy siltstone, concluding that water action causes internal damage to muddy siltstone samples, weakening the cementation between rock particles and leading to a decrease in uniaxial compressive strength and elastic modulus. Zhao Lizai et al. [15] conducted softening tests on the mechanical properties of water-saturated sandstone, concluding that the mechanical properties of sandstone gradually weaken with increased water saturation.

The aforementioned scholars primarily focus on studying the impact of water-rich conditions on tunnel damage to analyze corresponding prevention and control measures. They also conduct relevant physical experiments to examine the mechanical and acoustic emission characteristics of water-saturated rock samples, emphasizing the effects of submergence frequency and duration on the compressive and tensile strengths, modulus of elasticity, and the variations in acoustic emission characteristics of the rock samples [16–19]. In practical engineering applications, the perimeter rock of a tunnel can become submerged, and due to varying water levels and the inability to pump out water promptly [20–24], significant differences in the mechanical properties of the surrounding rock may arise.

Therefore, this paper systematically investigates the mechanical and acoustic emission characteristics of sandstone in circular roadway formations under varying depths of immersion. The research findings offer valuable insights for addressing practical challenges when the surrounding rock of a roadway is subjected to different water levels.

2. Experimental Equipment and Programs

2.1. Experimental Rock Sample Preparation

To maximize the geometric properties of the rock samples, the sandstone was machined into rectangular specimens measuring 100 mm × 100 mm × 25 mm, and their surfaces were polished to a smooth finish. The wave velocity of the sandstone was tested using an ultrasonic testing system, and several similar rock samples were selected. Using waterjet cutting technology, circular holes with a diameter of 12 mm were cut into the prefabricated rectangular rock samples to simulate circular channel morphology. The samples were categorized based on their depth of immersion, classified as bottom plate immersion, middle plate immersion, and top plate immersion. Additionally, the sandstone samples were numbered using the format 'letter-number'.

2.2. Test Equipment and Experimental Program

The control and acquisition of the test primarily involve a strain acquisition system, an acoustic emission acquisition system, and a test loading and control system. These components work in unison to effectively record the development of rock crack initiation points and manage fluctuations in the submerged water level. To achieve better control over the water level, the test loading and control system is designed with a closed tank. Additionally, a high-speed video camera is positioned to capture the entire experimental process, as illustrated in Figure 1. During the test, the three systems are synchronized to ensure consistent monitoring intervals, facilitating the processing and analysis of subsequent data [25,26].

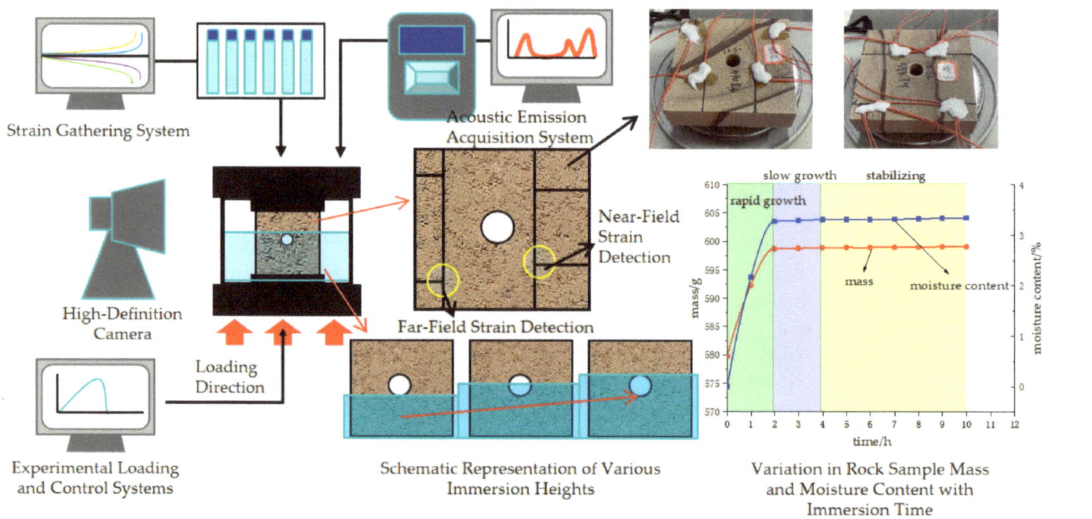

Figure 1. Experimental system.

To ensure that the specimen was in a dry state, it was placed in an oven at 105 °C for 12 h. Subsequently, the water content of the rock samples was monitored over time. The water content increased rapidly from 0 to 2 h, rose slowly from 2 to 4 h, and stabilized after 4 h. To guarantee that the samples were in a fully saturated state, they were immersed for 5 h. The curves depicting the changes in mass and water content of the rock samples during immersion are shown on the right side of Figure 1. Afterward, eight strain gauges were affixed to the surface of the rock sample in both axial and radial directions, positioned in near-field and far-field locations. Silica gel was applied on top to prevent the strain gauges from being affected by water. Additionally, two acoustic emission probes were attached to either side of the sample, and the effectiveness of the acoustic emission signal reception was evaluated using the broken lead method. At the beginning of the experiment, the loading and control system was set to a displacement rate of 0.002 mm/s. The acoustic emission acquisition system and the strain acquisition system were synchronized, and a high-definition video camera was positioned to record the rupture morphology of the rock samples. The four tasks are recorded simultaneously until the rock sample is completely fragmented during the recording process. The experimental flowchart is illustrated in Figure 2. Schematic diagrams depicting the immersion states of the specimens at various water depths are presented in the lower section of Figure 1. These diagrams represent three typical forms of immersion: bottom plate immersion, middle immersion, and top plate immersion.

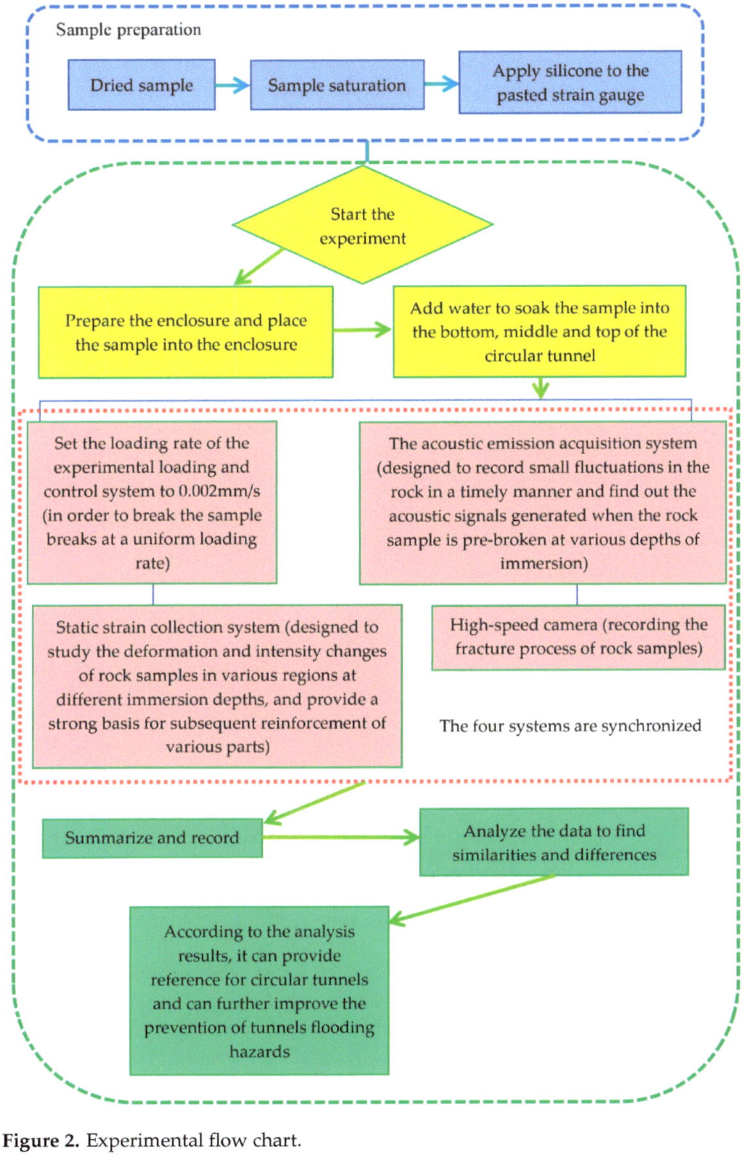

Figure 2. Experimental flow chart.

3. Results

3.1. Mechanical Characterization of Sandstone under the Influence of Submersion Depth

Sandstones often contain a class of minerals that are particularly susceptible to weakening due to water. When subjected to water infiltration, these minerals can alter the internal structure of the rock, impairing the bonding between particles and loosening the overall structure, which ultimately diminishes its mechanical performance. The softening of the rock becomes increasingly pronounced as the depth of water infiltration increases. To investigate the changes in compressive strength and deformation properties during this process—especially in rocks containing circular holes—a series of uniaxial compression experiments were designed and conducted, encompassing various water immersion depths. The primary objective of these experiments is to analyze how the uniaxial compressive strength values, modulus of elasticity, and other key physico-mechanical indices of sandstones with circular holes are affected by variations in water immersion depth. Based on the experimental data, we plotted the complete stress–strain relationship of the saturated rock samples under different immersion durations, as illustrated in Figure 3, thereby visualizing the dynamic relationship between these parameters [27–30].

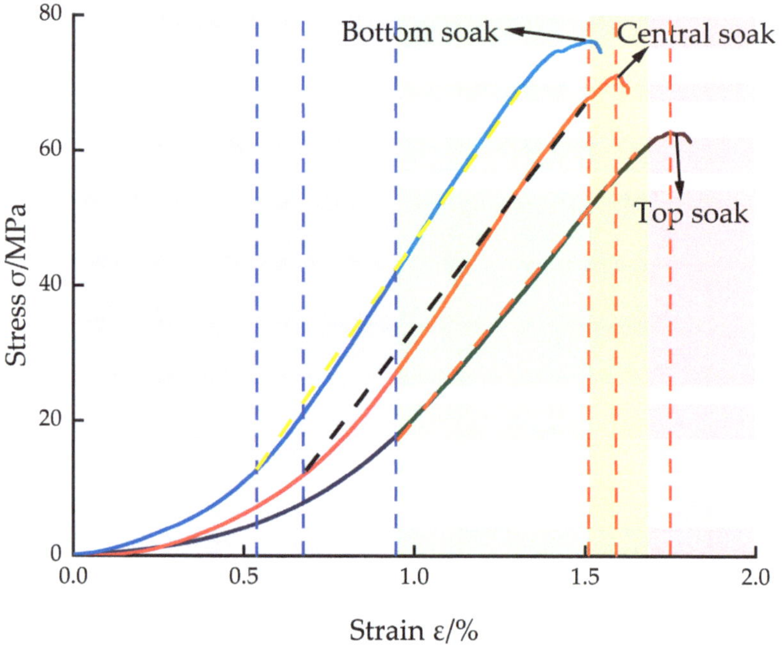

Figure 3. Stress−strain curves of sandstones with different immersion depths.

Figure 3 clearly illustrates that the depth of water immersion significantly influences the mechanical response of sandstones containing circular holes under uniaxial compression. As the duration of submergence increases, the trajectories of the complete stress–strain curves for these rock samples transition markedly from typical elastic deformation to nonlinear deformation characteristics. Specifically, under shallow submergence conditions, the sandstone with circular holes exhibits high uniaxial compressive strength and distinct brittle fracture characteristics during failure. This observation reveals that the internal structure of the rock still maintains a high degree of integrity and thus has the ability to withstand externally applied pressure. When the depth of submergence is increased to a moderate level, the curves display a notable downward section after reaching peak compressive strength, suggesting that the samples retain a certain degree of load-bearing capacity post-peak stress, although they are unable to resist further pressure. As the water immersion depth increases to a more significant level, a pronounced decline occurs after the peak compressive strength is achieved. This indicates that the rock samples have surpassed peak stress; while they still possess some load-bearing capacity, they exhibit more pronounced plastic deformation, with an increase in deformation compared to the case of bottom plate submergence, demonstrating stronger ductile damage characteristics. When the water immersion depth is further increased to the top plate, the slope of the ascending portion of the stress–strain curve decreases further, implying that the samples begin to accumulate greater deformation in the early stages of loading. After reaching peak strength, the curve does not drop sharply but instead undergoes a relatively gentle transition before gradually declining. This behavior indicates that the rock samples retain a considerable degree of residual strength after failure, with significantly enhanced deformation capacity and more pronounced toughness characteristics. This process clearly demonstrates that increased immersion depth facilitates the transformation of sandstone mechanical behavior from brittle to ductile.

The blue, green, and red areas in Figure 3 represent the compaction stage, elasticity stage, and yielding stage, respectively. The blue line segment indicates the onset of the elasticity stage for the rock samples at three different immersion depths, while the red line segment marks the conclusion of the yielding stage. Observations reveal that as the immersion depth of the rock samples increases, both the onset of the elasticity stage and the conclusion of the yielding stage shift progressively backward. However, the duration of the elasticity stage remains nearly constant across all three samples. This suggests that while increased immersion depth extends the compaction and yielding phases of the rock samples containing round holes, it has minimal impact on the duration of the elasticity phase.

Figure 4 illustrates the uniaxial compressive strength, maximum peak strain, and modulus of elasticity curves for sandstone rock samples containing circular holes at various immersion depths. The analysis reveals that the uniaxial compressive strength of sandstone is 76.29 MPa, with a peak strain of 1.52% under bottom plate immersion. Under middle plate immersion, the uniaxial compressive strength decreases to 71.04 MPa, representing a 6.88% reduction compared to the bottom plate immersion, while the peak strain increases to 1.60%, which is 8% higher than that of the sample under bottom plate immersion. Furthermore, the uniaxial compressive strength of sandstone under top plate immersion is 62.93 MPa, indicating an 11.42% decrease compared to the middle plate immersion, with a peak strain of 1.78%, which is 11.25% higher than that of the samples under middle plate immersion. As the immersion depth increases, both the uniaxial compressive strength and the elastic modulus of the sandstone with circular holes significantly decrease, while the peak strain gradually increases. Notably, the greater the immersion depth of the rock samples, the more rapidly the uniaxial compressive strength declines, and the peak strain correspondingly increases. This trend suggests that as the water immersion depth increases, the structure of the sandstone becomes progressively looser due to higher water saturation, leading to an increase in pore space and fissures. Consequently, the rock experiences greater deformation during compression before reaching the failure point. The increase in peak strain indicates that the rock can endure greater deformation prior to damage, despite a reduction in its load-bearing capacity (compressive strength). The water-saturated portion of the rock contributes to structural damage; as applied pressure increases, the formation of additional cracks and pores allows water to infiltrate, creating a 'lubrication' effect that reduces internal friction. This reduction in internal friction diminishes the cohesive properties of the rock, resulting in a decrease in the overall modulus of elasticity as immersion depth increases. This phenomenon is the primary cause of the observed decrease in the modulus of elasticity.

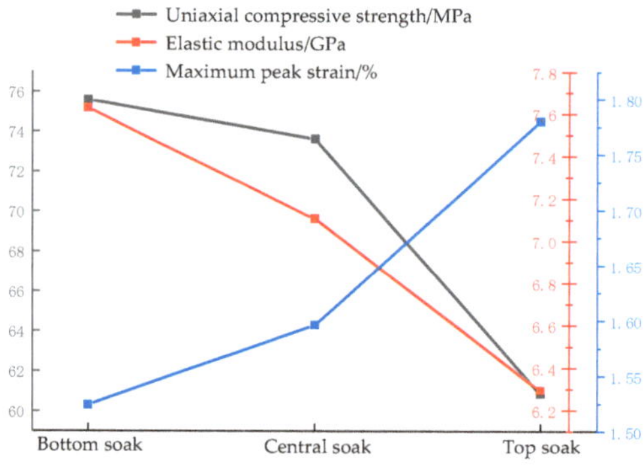

Figure 4. Relationship between different immersion depths and mechanical characteristics of rock samples.

3.2. *Characteristics of Sandstone Deformation under the Influence of Submergence Depth*

Under submerged water conditions, the deformation characteristics of sandstone rock samples containing round holes will vary significantly when subjected to the same force. This difference is attributed to the varying water depths in the roadway caused by different water levels. Underwater immersion, the initial fine cracks in the rock samples will change. As the immersion time increases, the microstructure of the rock samples will become looser, leading to damage in the bonding of the rock sample particles. Consequently, the mechanical properties will weaken, and the rock samples will gradually soften. By analyzing the change in rule between the radial and axial directions of the circular roadway in the bottom plate, top plate, and overall immersion at different water depths, a time–strain curve is plotted, as depicted in Figure 5 [31].

Circular channel morphology sandstone is under pressure in the depth of water to the bottom plate. According to Figure 5a, the maximum axial strain is 6%$^{-3}$, and the minimum is about 5%$^{-3}$. The far-field strain detection point, far from the center of the channel, detected a larger strain than the near-field detection point. The submerged part of the strain, both in the far-field and near-field, is greater than the non-immersed part of the strain. According to Figure 5b, it can be seen that among the radial strains under the subgrade immersion, the radial strain in the non-immersed part of the far-field is the largest at about 4.5%$^{-3}$, while the radial strain in the immersed part of the near-field is the smallest at about 1.5%$^{-3}$. Additionally, during the compression process, the strain of the non-immersed part increases more rapidly, resulting in a larger strain.

The sandstone exhibiting circular pore morphology is subjected to pressure at water depths up to the midpoint, as illustrated in Figure 5c. In this figure, the near-field non-immersed region experiences the highest strain, with a maximum axial strain of approximately 8%$^{-3}$ to 3%$^{-3}$. Conversely, the near-field immersed region shows the lowest strain, with an axial strain of about 1.5%$^{-3}$ to 3%$^{-3}$. The two far-field strain detection points, located farther from the circular channel, exhibit more similar strain values, both around 4%$^{-3}$ to 3%$^{-3}$. According to Figure 5d, the near-field immersed region has the largest radial strain, measuring approximately 3.5%$^{-3}$ to 3%$^{-3}$. The far-field non-immersed region displays the smallest axial strain, with a radial strain of about 1.2%$^{-3}$ to 3%$^{-3}$. Additionally, the far-field immersed region shows a radial strain similar to that of the far-field non-immersed region, indicating that the radial strains of both areas are quite comparable.

As shown in Figure 5e, the circular channelized morphology of the sandstone experienced complete submergence pressurization. The radial strain in the upper part of the near field is the largest, measuring approximately 7.5%$^{-3}$, whereas the axial strain in the lower part of the far field is the smallest, at around 1%$^{-3}$. According to Figure 5f, the radial strain in the upper part of the near field is the largest, with an axial strain of about 4%$^{-3}$, while the radial strain in the upper part of the far field is the smallest, with an axial strain of about 1%$^{-3}$.

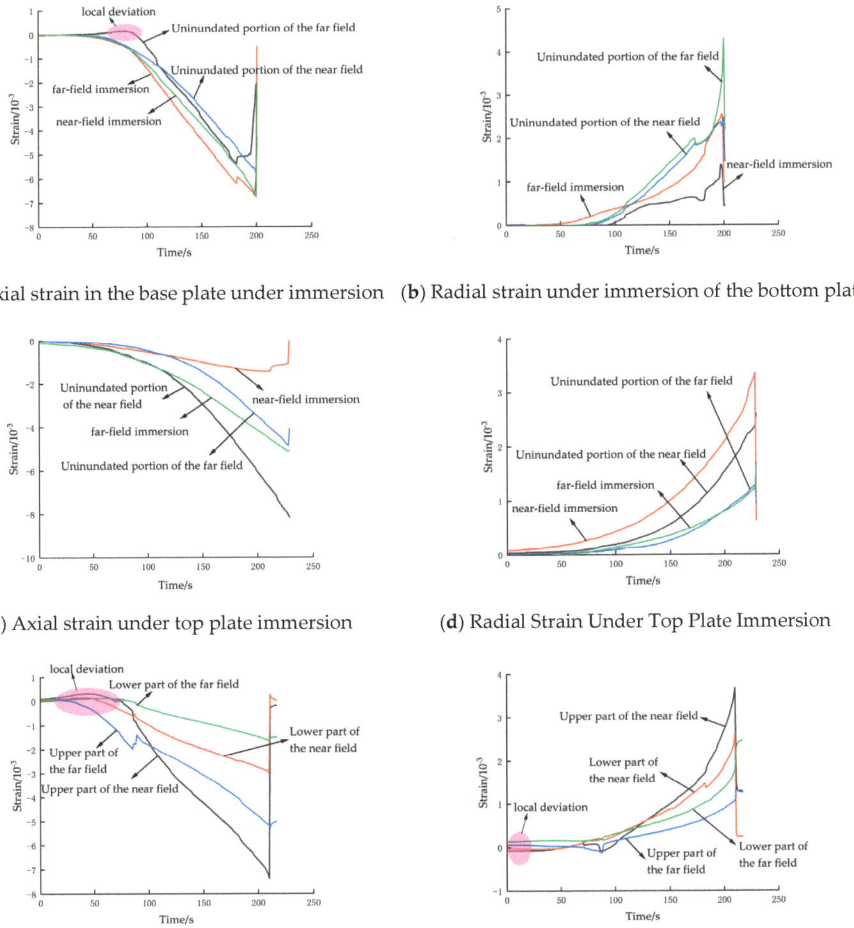

(**a**) Axial strain in the base plate under immersion (**b**) Radial strain under immersion of the bottom plate

(**c**) Axial strain under top plate immersion (**d**) Radial Strain Under Top Plate Immersion

(**e**) Axial strain under full immersion (**f**) Radial strain under full immersion

Figure 5. Time–strain curves under different water immersion depths.

Through the comprehensive analysis of the time–strain curves at various immersion depths, it is observed that the strain of the sandstone specimens with rounded channel morphology deviates locally during bottom immersion and full immersion. This deviation may be attributed to the pressurization of the specimen under water, causing slight displacement of the monitoring point initially. The strain curves of the rock samples under three different water depth immersions are consistent with the fissure evolution pattern of the rock samples. The uniaxial compressive tests conducted on the rock samples under water immersion conditions revealed that the axial strain was consistently greater than the radial strain, regardless of the level of water immersion. The deformation curves of the rock samples were smoother during the compression-tightening stage. Upon entering the elastic stage, the friction between the cracks of the rock samples decreased as a result of the significant immersion in water. The strains were influenced by the water, causing irregular fluctuations in the axial and radial strain curves. Water alters the pore structure and permeability of the rock, resulting in modifications to the rock's deformation properties [32]. Increasing the depth of immersion will make the deformation of the rock samples more significant.

3.3. Characteristics of Sandstone Damage under Different Immersion Depths

In order to analyze the influence mechanism of crack propagation in sandstone containing circular holes under submerged water conditions during the rock damage process, the crack extension paths and mechanisms of rock samples were examined by extracting images of crack propagation at various submerged water depths.

To better analyze the effects of submerged water conditions on the damage to sandstone with circular holes, the diagrams are sequentially marked to indicate the order of crack generation and the type of crack (tensile or shear) for quantitative analysis and assessment. The letters in Figure 6 denote the types of cracks: A represents a shear crack, which typically extends from the edges of the holes in the rock samples and propagates toward the edges of the samples, with the direction of propagation being diagonal, either upward or downward. B denotes a tensile crack, which occurs when the rock samples experience tensile stresses in both upward and downward directions, usually extending along the axial direction. This tensile crack may deviate to some extent when subjected to various external conditions [33–36]. The angular numbers indicate the chronological order of crack generation. The red line segments represent shear cracks, the blue line segments represent tensile cracks, and the dashed lines indicate secondary cracks formed during crack propagation.

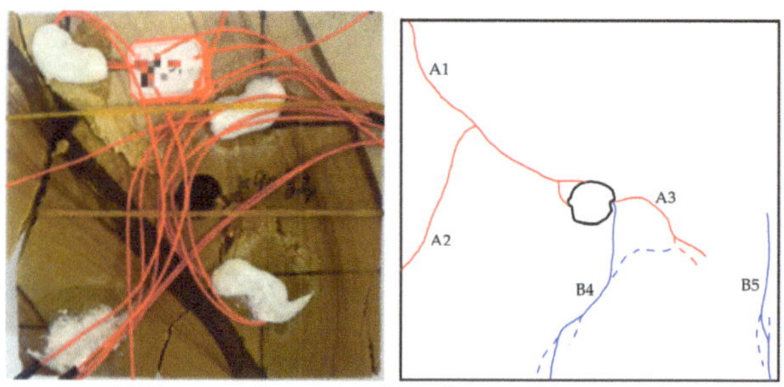

(**a**) Damage diagram under submerged bottom plate

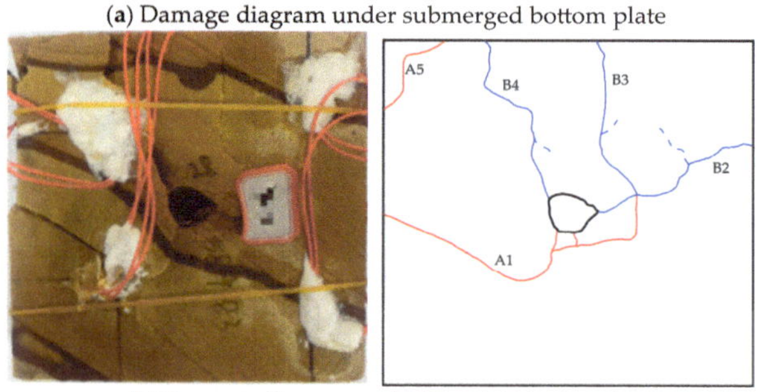

(**b**) Damage diagram under top plate immersion

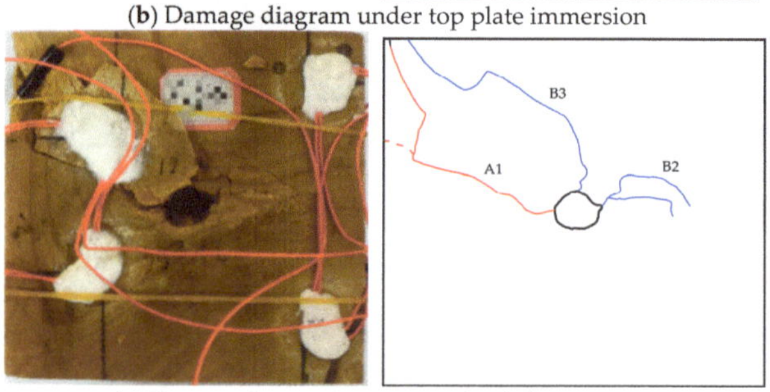

(**c**) Damage destruction map under full immersion

Figure 6. Damage and failure diagram of rock samples under immersion at different depths.

Under submerged water conditions at the base plate, the damage diagram of the rock sample is illustrated in Figure 6a. During the process of applying axial gradual pressure, a shear crack A1 emerged from the left side of the hole, extending toward the upper left. As the crack propagated, a second shear crack A2 developed, extending toward the lower left, followed by another shear crack A3 that formed on the right side of the hole, extending toward the lower right. With the instability of the upper part of the sample, new cracks formed in the area below the hole, including two tension cracks B4 and B5 that extended from the top to the bottom, along with several secondary cracks. The damage schematic diagram of the rock sample, which features a circular hole morphology in sandstone under central submerged water conditions, is presented in Figure 6b. Initially, a shear crack A1 formed in the region below the hole. As the loading pressure gradually increased, a tensile crack B2 appeared, extending to the upper right from the right side of the hole. Subsequently, a second tensile crack B3 developed, extending upward during the propagation process. Additionally, new cracks B4 and B5 extended to the upper left from the right side of the hole, accompanied by several secondary cracks originating from the upper side of the hole. A tensile crack B4 also extended to the upper left, and the final shear crack A5 was generated in the last stage of the sample rupture. The damage schematic of the rock sample, featuring a circular hole morphology in sandstone under top plate flooding conditions, is depicted in Figure 6c. A total of three main cracks were generated, all extending from the holes to various regions of the rock sample. The first crack, a shear crack A1, extended from the left to the upper left. This was followed by two tensile cracks: B2, extending to the upper right, and B3, extending to the upper left in the upper part of the hole. All three cracks were relatively concentrated in the upper part of the rock sample, which remained unsaturated.

The analysis was conducted at three different submergence depths, with pressure applied at the water depth submerged to the bottom plate. Water infiltrated the rock body, resulting in water-induced splitting and fracturing below. The rock samples surrounding the holes experienced local softening, which adversely affected their overall compressive strength. When pressurized at the water depth submerged to the middle, water penetrated the rock body to a greater extent. The portion of the rock that remained above the water was significantly more brittle than the submerged section, resulting in the formation of smaller cracks. However, the primary cracks leading to overall structural imbalance were still located in the submerged area, indicating that the water-saturated portion is more susceptible to damage than the dry part. In the case of complete submergence, the rock was entirely saturated with water. This water absorption led to an increase in the volume of the rock samples, which in turn raised the internal stress and made the formation of tensile cracks within the rock samples more likely. It is evident that as the submergence depth increases, the rock samples become increasingly prone to tensile cracks, while the number of secondary cracks that develop during the crack propagation process gradually decreases.

To thoroughly investigate the extension characteristics of cracks at various developmental stages, we utilized the RA (rise time/amplitude) and AF (average frequency) values from the acoustic emission technique as analytical tools to systematically quantify and analyze the damage patterns of cracks. We selected several sandstone samples exhibiting circular pore characteristics for the immersion test and processed the acoustic emission data points collected during the experiment. Subsequently, we plotted these data points on RA-AF diagrams for visual representation [37,38]. Based on the distribution characteristics of the acoustic emission signals under two distinct rupture mechanisms—tension and shear—we clearly delineated a dividing line on the RA-AF diagrams to differentiate between these two damage modes. Furthermore, we calculated the slope of this demarcation line to be 5.48, a result clearly illustrated in Figure 7. Through this analysis, we gained a more precise understanding of the crack extension mechanism and its characteristics at different stages.

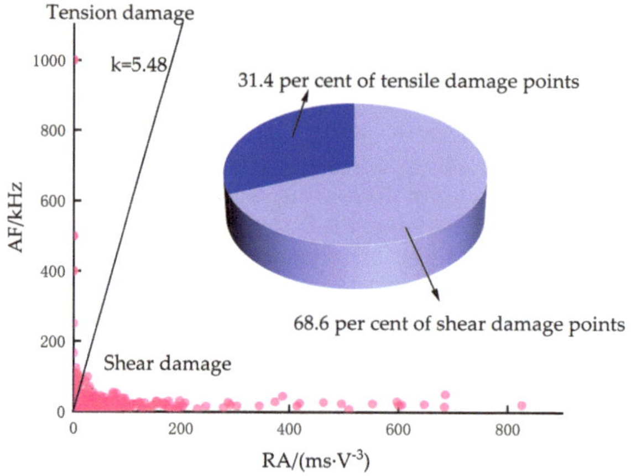

(**a**) Bottom soak

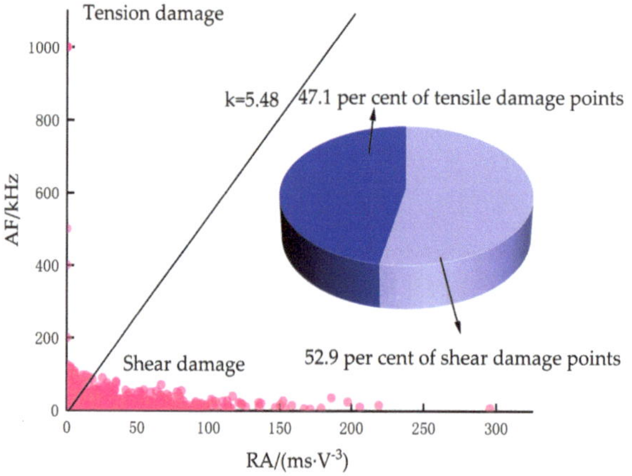

(**b**) Central soak

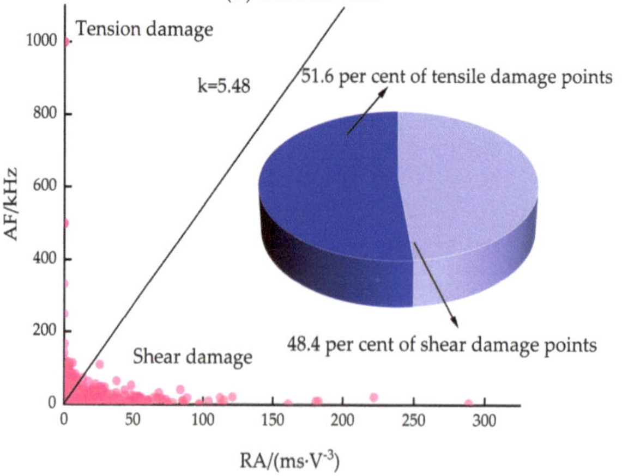

(**c**) Top soak

Figure 7. Distribution of acoustic emission RA−AF at different immersion depths.

From the analysis of the percentages of crack types, it is evident that under bottom plate immersion, the percentage of tensile damage points is 31.4%, which is significantly lower than the percentage of shear damage points. Under middle plate immersion, the percentage of tensile damage points rises to 47.1%, approaching the percentage of shear damage points. In contrast, under top plate immersion, the percentage of tensile damage points increases to 51.6%, surpassing the percentage of shear damage points. By comparing the damage patterns of rock samples at different immersion depths, as illustrated in Figure 7, shear and tensile cracks can be clearly distinguished. Observations of each type of damage point indicate that as immersion depth increases, the tensile damage in sandstone rock samples containing round holes progressively intensifies.

Some scholars have reached a consensus that the proportion of tensile cracks in rocks tends to increase as the height of water ingress rises. However, through careful observation and analysis presented in this paper, we have identified a more complex phenomenon: while the number of tensile cracks does increase with greater immersion depth, the growth rate does not continuously accelerate. Instead, it exhibits a trend of gradual decline as immersion depth increases. This finding indicates that the relationship between immersion depth and the expansion rate of tensile cracks is not a simple linear correlation. Rather, it suggests the involvement of more intricate physical and chemical mechanisms, which may include the gradual adaptation of the rock's internal structure, the osmotic pressure effects of water, and the mechanical properties of the rock material itself. Therefore, future studies should further investigate these mechanisms to achieve a more comprehensive understanding of how immersion depth influences the behavior of tensile crack propagation in rocks.

3.4. Acoustic Emission Characteristics of Sandstone under Different Immersion Depths

During the fracture process, rock accumulates internal strain energy, which is manifested through complex transformations and intermittent releases. The acoustic emission monitoring system can detect the acoustic emission signals generated during the rock damage process. This signal information accurately reflects the expansion of cracks and the extent of damage within the rock. The acoustic emission characteristics of sandstone containing circular holes also exhibit variations at different immersion depths. Acoustic emission ringing counts and cumulative energy are selected as parameters to analyze the intrinsic relationship between the acoustic emission characteristics of sandstone with circular holes and rock damage under varying immersion depths [39–42]. To ensure that the overall characteristics are accurately represented, the most representative test results under different immersion depth conditions are selected for analysis. Based on the test data, time–stress–acoustic emission ringing counts–cumulative energy curves are plotted, as shown in Figure 8. To better illustrate the relationship between immersion depth and acoustic emission phenomena, the curves are divided into four phases: the compression-tightening phase, the elasticity phase, the yielding phase, and the post-peak stage, allowing for a detailed analysis of their main features.

It can be observed from Figure 8 that all water-immersed rock samples exhibit acoustic emission phenomena to varying degrees throughout all stages of the pressurized damage process. The maximum acoustic emission ringing counts for all samples occur near the peak stress; however, notable differences are evident among the samples at varying depths of immersion across the different stages.

In the OA stage (the compaction stage), purple part, a small amount of acoustic emission phenomena occurs in both bottom-immersed and middle-immersed circular pore-bearing sandstones, while acoustic emission signals from top-immersed circular pore-bearing sandstones are almost nonexistent. The acoustic emission ringing frequency of sandstones under bottom-flooded conditions is approximately one to three times greater than that of sandstones flooded in the center and six to nine times greater than that of sandstones flooded at the top. As the duration of submergence increases, the acoustic emission phenomena of sandstones under the three different submergence depth conditions decrease to varying degrees. This reduction can be primarily attributed to the relatively low stress levels in the rock samples during the compression and densification stage. When pressure is applied, the original small cracks within the rock samples begin to close. During this closure process, damage occurs between the rock cracks, and some friction between rough particles generates a small amount of acoustic emission. Additionally, the presence of water leads to the immersion of cracks, resulting in the softening of the rock. Water also reduces sound propagation, which contributes to the lower acoustic emission ringing counts. As immersion time increases, the softening effect of water on the rock intensifies, further decreasing the acoustic emission phenomena.

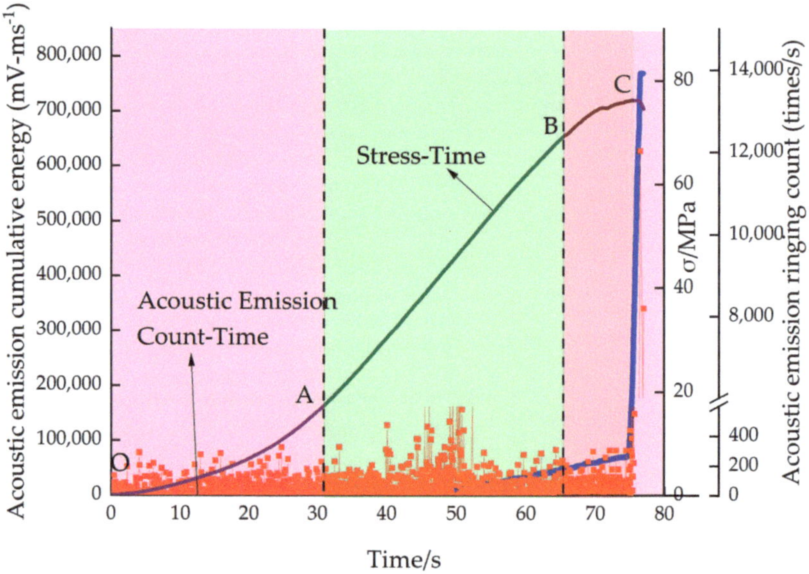

(**a**) Bottom soak

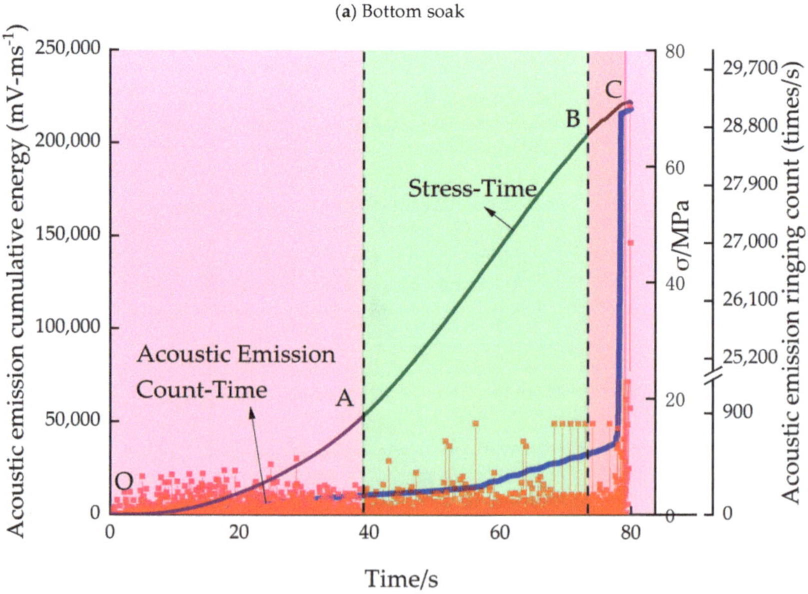

(**b**) Central soak

Figure 8. *Cont.*

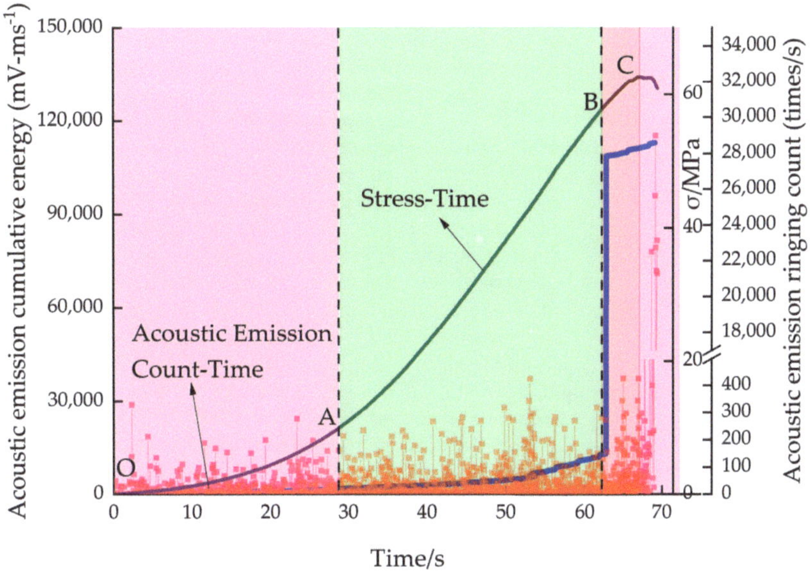

(**c**) Top soak

Figure 8. Acoustic emission ringing count—stress—time curve of rock samples.

The AB stage, green part, also known as the elasticity stage, is characterized by minimal acoustic emission phenomena in the rock samples situated between the middle and top plates during water immersion. This stability arises because, at this stage, the rock primarily undergoes elastic deformation, with minimal plastic deformation occurring. As a result, the generation of acoustic emission signals is limited. While the rock samples at the bottom plate also tend to stabilize during water immersion, slight fluctuations may occur in the middle section. These fluctuations could be attributed to minor detachment of the surface rock layer in the circular hole sandstone, which may generate some acoustic signals.

In the BC stage (yielding stage), yellowish part, damage accumulation contains precursor information indicative of instability. It is observed that during the yielding stage, the acoustic emission counts of the rock samples decreased significantly with increasing immersion depth. Slight fluctuations in the acoustic emission ringing counts of the middle submerged rock samples suggest that crack initiation and expansion during this phase contributed to the accumulation of acoustic emission ringing counts. Additionally, the acoustic emission signals decreased with prolonged submergence time, indicating that the softening effect of water on the rock samples suppressed the generation of these signals. At both the bottom and top slab immersion depths, the acoustic emission signals of the rock samples did not fluctuate significantly. This is likely because this stage represents the initial phase of energy accumulation, where the stress and energy within the rock are insufficient to reach the critical threshold for rupture or significant deformation, resulting in minimal acoustic emission activity. Furthermore, under submerged conditions, water infiltrates the pores and fissures of the sandstone, providing a buffering effect that slows the concentration of stress. Consequently, this reduces stress concentration and crack propagation during the compaction stage, leading to a decrease in the generation of acoustic emission signals.

The post-peak stage, which follows point C, lilac part, exhibits more pronounced acoustic emission signals in sandstones with circular pore morphology at varying submergence depths. Following the destabilization and damage of the rock, a significant amount of energy accumulated during the previous stage is released instantaneously. This release, along with localized damage and particle spalling within the rock, generates additional acoustic emission signals. Under submerged conditions, water acts as a lubricant for crack propagation, facilitating the extension of cracks and resulting in more frequent localized damage. Stress redistribution occurs within the rock, as stresses that were initially concentrated in specific areas are released and transferred to other regions, leading to the initiation of new crack extensions and further damage. This dynamic alteration in stress levels produces a substantial increase in acoustic emission signals, which are significantly higher than those observed during the pressure-tight, elastic, and yield stages.

The cumulative acoustic emission energy of sandstone containing circular roadway holes at various immersion depths is illustrated in Figure 8. It is evident that the total cumulative acoustic emission energy of the rock samples decreases as the immersion depth increases. Notably, the total energy of the rock samples subjected to deep immersion in the bottom plate water is the highest, measuring 742,547, while the total energy of the rock samples under full immersion is the lowest, at 113,172.

Taking the above analyses into account, we can conclude that the change in the acoustic emission ringing counts demonstrates a high degree of consistency with the damage evolution law within the rock samples. Under submerged conditions, the inherent damage of the rock samples significantly influences the fluctuations in the acoustic emission signal as pressurization time progresses. Additionally, the softening and suppressive effects of water also play a crucial role in these fluctuations. Although the depth of immersion has a complex impact on the acoustic emission signals of sandstone specimens with circular pore morphology, a general trend emerges: the cumulative energy of acoustic emission decreases as the depth of immersion increases.

The study conducted by Zihui Zhu [43] and other researchers found that in a dry state, the acoustic emission energy curve typically progresses through three phases: smooth, slow rise, and sudden rise. However, as water content increases, the distribution characteristics of these three stages change significantly. Notably, in a water-saturated state, the acoustic emission energy curve is nearly simplified to just two phases: a smooth increase and a sudden increase. By examining the acoustic emission curves under submerged conditions, we also observe that variations in immersion depth influence the pattern of acoustic emission activity. Specifically, during the 'smooth' phase, the number of acoustic emission events decreases as immersion depth increases, resulting in smoother behavior during this phase. These findings provide a crucial experimental foundation for a deeper understanding of the mechanical behavior and damage evolution of rocks under submerged conditions, revealing the mechanisms by which submergence depth and water content affect the acoustic emission activity of rocks.

4. Discussion

(1) The experimental results indicate a significant influence of immersion depth on the mechanical properties of sandstone with a circular pore morphology. Both the uniaxial compressive strength and the modulus of elasticity of the sandstone decreased markedly as the water immersion depth increased. This reduction may be attributed to the rise in pore water pressure within the rock samples caused by the immersion, which subsequently diminishes the bearing capacity of the samples. Additionally, the gradual increase in peak strain may suggest an enhanced capacity for plastic deformation in the rock samples under conditions of water immersion.

(2) The influence of water infiltration on strain monitoring should not be overlooked, as it can lead to deviations in strain data, thereby affecting the accurate assessment of the mechanical properties of rock samples. Consequently, the impact of water intrusion on monitoring data must be thoroughly considered when conducting relevant experiments, and appropriate measures should be implemented to correct the data.

(3) The observation of crack extension provides crucial insights into the damage mechanisms of sandstone under water immersion conditions. The percentage of tension damage points gradually increased with greater immersion depth, likely due to the reduced friction between cracks within the rock samples caused by water immersion, which facilitates the formation of

tension cracks. Additionally, the decrease in the number of secondary cracks may indicate a gradual homogenization of the damage pattern in the rock samples under these conditions.

(4) The reduction in cumulative acoustic emissions from the rock samples, along with its correlation to the development of internal damage, further substantiates the impact of immersion depth on the mechanical properties of sandstones. The decline in total cumulative acoustic emission energy as immersion depth increases may be attributed to the gradual accumulation and progression of internal damage within the rock samples. Consequently, the effect of immersion depth on the mechanical properties of sandstone must be thoroughly considered in related research and engineering applications to ensure the safety and stability of projects.

5. Conclusions

(1) The uniaxial compressive test of rock samples with round holes at varying immersion depths reveals several key findings. The uniaxial compressive strength of sandstone under bottom plate immersion is 76.29 MPa, with a peak strain of 1.52%. Under middle plate immersion, the uniaxial compressive strength of the sandstone decreases to 71.04 MPa, which is 6.88% lower than that of the samples under bottom plate immersion. However, the peak strain increases to 1.60%, representing an 8% increase compared to the bottom plate immersion samples. For sandstone under top plate immersion, the uniaxial compressive strength further declines to 62.93 MPa, which is 11.42% lower than that of the middle plate immersion samples. The peak strain in this case rises to 1.78%, indicating an 11.25% increase over the middle immersion samples. Overall, as immersion depth increases, both the uniaxial compressive strength and elastic modulus of the sandstone with round holes significantly decrease, while the peak strain gradually increases. Notably, the greater the immersion depth of the rock sample, the more rapidly its uniaxial compressive strength diminishes, and the more quickly the peak strain escalates. The depth of immersion leads to elongation of the compaction and yield phases of samples containing circular holes, but does not have an excessive effect on the duration of the elastic phase.

(2) When sandstone rock samples are pressurized in water at different immersion depths, water percolation leads to partial deviations in strain monitoring. The strain curves of the rock samples under three different water immersion depths are consistent with the fissure evolution pattern of the rock samples. The uniaxial compressive tests on rock samples under water immersion conditions showed that the deformation curves of the rock samples were relatively smooth in the compression-tight stage, and when entering the elasticity stage, the friction between the cracks of the rock samples was reduced due to the extensive immersion of the water body. The strains were affected by the water body, which led to irregular fluctuations in the axial and radial strain curves.

(3) The crack extension of the rock samples containing round holes is 31.4% of the tensile damage points in the bottom plate, which is much lower than the percentage of shear damage points; it is 47.1% of the tensile damage points in the middle plate, which is close to the percentage of shear damage points, and 51.6% of the tensile damage points in the top plate, which is higher than the percentage of shear damage points. With the increase of water immersion depth, the tensile damage points of sandstone with circular hole pattern also increase, which is more likely to produce tensile cracks, and with the gradual increase of water immersion depth, the number of secondary cracks will be gradually reduced.

(4) The cumulative number of acoustic emissions of the rock samples decreases with the increase of water immersion depth. The curve of the cumulative number of acoustic emissions reflects the stage-by-stage law of the evolution of internal damage of the rock samples at different immersion times, which has a consistent correspondence with the degree of internal damage of rock samples, and the total cumulative energy of acoustic emissions of the rock samples decreases with the increase of the depth of the water level.

(5) This study primarily focused on sandstone rock samples exhibiting circular pore morphology and did not consider rocks with other shapes, materials, or structures; therefore, the generalizability of the results may be limited. Additionally, the experimental

conditions (e.g., immersion rate, temperature, and water quality) may influence the results, but these factors were not thoroughly investigated or controlled in this study. While the reduction in the number of acoustic emission accumulations and energy corresponds to the degree of internal damage in the rock samples, it does not provide a comprehensive exploration of the specific relationship between the acoustic emission signals and the changes in the rock microstructure. Furthermore, it does not address other factors (e.g., noise interference) that may impact the acoustic emission signals.

Author Contributions: G.L. and S.W. wrote the main text of the manuscript and the experimental data processing and summaries, while Y.Z., D.W. and Z.Y. were the co-authors responsible for collecting relevant information to organize the structure of the paper. All authors have read and agreed to the published version of the manuscript.

Funding: This research was funded by Natural Science Foundation of Heilongjiang Province, grant number YQ2023E039; Basic Scientific Research Operating Expenses of Heilongjiang Provincial Universities and Colleges of China, grant number 2022-KYYWF-0554; Scientific and Technological Key Project of "Revealing the List and Taking Command" in Heilongjiang Province, grant number 2021ZXJ02A03, 2021ZXJ02A04.

Data Availability Statement: The datasets used and analyzed in this study can be provided by the authors Gang Liu and Shengxuan Wang according to reasonable requirements.

Acknowledgments: This research was supported by the Heilongjiang Ground Pressure and Gas Control in Deep Mining Key Laboratory, Heilongjiang University of Science and Technology.

Conflicts of Interest: Authors Gang Liu and Zhitao Yang were employed by the company Baotailong New Material Co., Ltd. The remaining authors declare that the research was conducted in the absence of any commercial or financial relationships that could be construed as a potential conflict of interest.

References

1. Chen, X.-J.; Fang, P.-P.; Chen, Q.-N.; Hu, J.; Yao, K.; Liu, Y. Influence of cutterhead opening ratio on soil arching effect and face stability during tunnelling through non-uniform soils. *Undergr. Space* **2024**, *17*, 45–59. [CrossRef]
2. Gu, H.; Lai, X.; Tao, M.; Cao, W.; Yang, Z. The role of porosity in the dynamic disturbance resistance of water-saturated coal. *Int. J. Rock Mech. Min. Sci.* **2023**, *166*, 105388. [CrossRef]
3. Zhao, Z.; Yang, P.; Zhang, M.; Lv, X.; Chen, S. Stability analysis of weakly cemented soft rock tunnel enclosure under the combined effect of water environment and non-uniform ground pressure. *J. Min. Saf. Eng.* **2022**, *39*, 126–135. [CrossRef]
4. Li, B.; Li, X.; Ren, S. Analysis and control of coal seam roadway instability under complex water environment. *J. Min. Saf. Eng.* **2011**, *28*, 370–375.
5. Wu, B.-Y.; Wang, W.-N.; Guo, D.-M. Strength damage and acoustic emission characteristics of fissured sandstone under the influence of the number of immersion. *J. Min. Saf. Eng.* **2020**, *37*, 1054–1060. [CrossRef]
6. Xia, D.; Yang, T.; Xu, T.; Wang, P.; Zhao, Y. Experiments on the effect of immersion time on the acoustic emission characteristics of water-saturated rocks during damage destruction. *J. Coal* **2015**, *40*, 337–345. [CrossRef]
7. Xia, D.; Yuan, X. Experimental study on the effect of soaking time on shear strength parameters of water-saturated rocks. *Min. Res. Dev.* **2015**, *35*, 65–69. [CrossRef]
8. Xia, D.; Chang, H.; Lu, H.; Yuan, X. Experimental study on the effect of immersion time on longitudinal wave velocity in water-saturated rocks. *Min. Res. Dev.* **2016**, *36*, 68–71. [CrossRef]
9. Jiang, L.; Zhao, D.; Zhou, J. Research on acoustic emission characteristics of shale with different water content under uniaxial compression test. *J. Undergr. Space Eng.* **2021**, *17*, 696–702.
10. Tan, T.; Zhao, Y. Tests on mechanical properties of sandstone under dry and water-saturated condition. *Min. Eng. Res.* **2022**, *37*, 15–23. [CrossRef]
11. Wang, Y.; Liu, X.; Wang, L.; Li, Z.; Chen, Y.; Rong, T. Effects of loading rate and water saturation on mechanical behaviour and microscopic damage characteristics of sandstone. *J. Min. Saf. Eng.* **2022**, *39*, 421–428. [CrossRef]
12. Xiao, F.; Gao, Y. Mechanical properties of dry and water-saturated pore rock samples under cyclic loading. *J. Heilongjiang Univ. Sci. Technol.* **2023**, *33*, 673–681.
13. Yin, D.; Ding, Y.; Wang, F.; Jiang, N.; Tan, Y.; Li, Z. Experimental study on tensile properties of coal rock under pressure water immersion. *J. Rock Mech. Eng.* **2023**, *42*, 3178–3191. [CrossRef]
14. Fang, J.; Yao, Q.L.; Wang, W.N.; Tang, C.; Wang, G. Study on the effect of water content on strength damage and acoustic emission characteristics of muddy siltstone. *J. Coal* **2018**, *43*, 412–419. [CrossRef]
15. Zhao, L.C. Experimental study on the softening of mechanical properties of water-saturated sandstone. *J. Undergr. Space Eng.* **2022**, *18*, 154–162.

16. Yaghoubi, A.A.; Zeinali, M. Determination of magnitude and orientation of the in-situ stress from borehole breakout and effect of pore pressure on borehole stability—Case study in Cheshmeh Khush oil field of Iran. *J. Pet. Sci. Eng.* **2009**, *67*, 116–126. [CrossRef]
17. Yu, C.Y.; Tang, S.B.; Tang, C.A.; Duan, D.; Zhang, Y.J.; Liang, Z.Z.; Ma, K.; Ma, T.H. The effect of water on the creep behavior of red sandstone. *Eng. Geol.* **2019**, *253*, 64–74. [CrossRef]
18. Zhang, H.; Yin, S.D.; Aadnoy, B. Poroelastic modeling of borehole break-outs for in-situ stress determination by finite element method. *J. Pet. Sci. Eng.* **2018**, *162*, 674–684. [CrossRef]
19. Yu, C. Experimental and Numerical Simulation Study on the Effect of Water on the Mechanical Properties of Rocks. Ph.D. Thesis, Dalian University of Technology, Dalian, China, 2020. [CrossRef]
20. Chen, Z.; Guo, N. New developments of mechanics and application for unsaturated soils and special soils. *Geotechnical* **2019**, *40*, 1–54. [CrossRef]
21. Cui, H.; Tang, J.; Jiang, X. Experimental study on physical mechanics and roughness of splitting surface of high-temperature granite after combined cooling. *J. Rock Mech. Eng.* **2021**, *40*, 1444–1459. [CrossRef]
22. Liu, G.; Li, Y.; Xiao, F.; Huang, S.; Zhang, R. Study on failure mechanical behavior and damage evolution of yellow sandstone under uniaxial, triaxial and pore water. *J. Rock Mech. Eng.* **2019**, *38*, 3532–3544. [CrossRef]
23. Liu, G.; Xiao, F.; Qin, T. Small size effect of rock mechanics features and acoustic emission rules. *J. Rock Mech. Eng.* **2018**, *37*, 3905–3917.
24. Liu, G.; Wang, D.; Zan, Y.; Wang, S.; Zhang, Q. Feasibility Study of Material Deformation and Similarity of Spatial Characteristics of Standard Coal Rocks. *Processes* **2024**, *12*, 454. [CrossRef]
25. Mastin, L.G. Development of Borehole Breakouts in Sandstone. Master's Thesis, Stanford University, Berkeley, CA, USA, 1984.
26. Mao, D.W.; Du, S.H.; Li, D.Y.; Shi, B.; Ruan, B. Mechanical behavior and submerged wetting of altered granite based on large-scale triaxial tests. *J. Rock Mech. Eng.* **2020**, *39*, 1819–1831. [CrossRef]
27. Xiao, Y.M.; Qiao, Y.F.; He, M.C.; Li, A.G. Strain rate effect of silt-stone under triaxial compression and its interpretation from damage mechanics. *Rock Mech. Rock Eng.* **2023**, *56*, 8643–8656. [CrossRef]
28. Wong, L.; Maruvanchery, V.; Liu, G. Water effects on rock strength and stiffness degradation. *Acta Geotech.* **2016**, *11*, 713–737. [CrossRef]
29. Liu, H.; Yang, C.; Liu, X.; Zhu, W. Study on the submerged shear creep characteristics and the intrinsic modelling of rock structural surfaces. *Met. Min.* **2022**, 71–80. [CrossRef]
30. Luo, C. Influence of Tailings Water Leaching Action on Macro and Fine Structure and Mechanical Characteristics of Sandstone. Master's Thesis, Liaoning University of Science and Technology, Anshan, China, 2023. [CrossRef]
31. Feng, G.; Yu, F.; Dai, Z.; Chen, S.; Li, J. Experimental study of water absorption and expansion aging characteristics of red bedded mudstone in central Sichuan. *J. Rock Mech. Eng.* **2022**, *41*, 2780–2790. [CrossRef]
32. Zhou, Q.; Ma, D.; Deng, R.; Kang, J.; Zhu, Q. Experimental study on mechanical properties of red bed soft rock under the action of geothermal system. *Geotechnics* **2020**, *41*, 3333–3342. [CrossRef]
33. Wei, D.; Gu, H.; Wang, C.; Wang, H.; Zhu, H.; Guo, Y. Extension Mechanism of Water-Conducting Cracks in the Thick and Hard Overlying Strata of Coal Mining Face. *Water* **2024**, *16*, 1883. [CrossRef]
34. Zhou, Z.L.; Cai, X.; Ma, D.; Du, X.M.; Chen, L.; Wang, H.Q.; Zang, H.Z. Water saturation effects on dynamic fracture behavior of sandstone. *Int. J. Rock Mech. Min Sci.* **2019**, *114*, 46–61. [CrossRef]
35. Zhang, X.; Peng, C.; Liu, C. Study on crack propagation and energy evolution characteristics of water-saturated red sandstone containing hole [J/OL]. *J. Civ. Environ. Eng.* **2024**, 1–8. Available online: http://kns.cnki.net/kcms/detail/50.1218.TU.20240711.1723.004.html (accessed on 2 September 2024). (In English and Chinese)
36. Fang, H. Research on Deformation and Damage Characteristics and Control Mechanism of Roadway Perimeter Rock in Water-rich Coal Seam. Ph.D. Thesis, China University of Mining and Technology, Beijing, China, 2021. [CrossRef]
37. Chen, T.; Yao, Q.; Du, M.; Zhu, C.; Zhang, B. Experimental study on the damage of fissure development of coal samples by the number of immersion. *J. Rock Mech. Eng.* **2016**, *35*, 3756–3762. [CrossRef]
38. Zhong, Z.; Li, A.; Deng, R.; Wu, P.; Xu, J. Experimental study on the time-expansion and deformation characteristics of red-bedded mudstone in central Sichuan. *J. Rock Mech. Eng.* **2019**, *38*, 76–86. [CrossRef]
39. Luo, K.; Bai, Y.; Zhao, S.; Xia, D.; Wu, C.; Jia, Y. Experimental study on the effect of water saturation time on the acoustic emission characteristics of rocks containing clay minerals. *Min. Eng. Res.* **2022**, *37*, 26–36. [CrossRef]
40. Ma, Y.; Li, X.; Zhai, S.; Huang, Q. The true triaxial test of strain field and acoustic emission response Bearing failure of coal with prefabricated borehole. *J. China Univ. Min. Technol.* **2024**, *53*, 497–508. [CrossRef]
41. Guo, H.; Duan, B.; Wang, W. Acoustic emission characteristics of sandstone with different leaching heights under tailings water erosion. *Hunan Nonferrous Met.* **2024**, *40*, 79–82.
42. Chen, G.; Zhang, J.; Li, T.; Chen, S.; Zhang, G.; Lv, P.; Teng, P. Mechanical property damage deterioration mechanism of coal-rock assemblage under water-rock interaction. *J. Coal* **2021**, *46*, 701–712. [CrossRef]
43. Zhu, Z.; Guo, J.; Sun, F.; Zhang, H. Experimental study of acoustic emission and crack extension in fissured sandstone under different water-bearing states. *J. High Press. Phys.* **2023**, *37*, 91–101.

water

MDPI

Article

Experimental Study for the Matching of Explosives and Rocks Based on Rock Hydrophysical Properties

Zhaozhen Zhu and Zhiyong Zhou *

School of Resources and Safety Engineering, Central South University, Changsha 410083, China; 15367495493@163.com

* Correspondence: csuzzy@csu.edu.cn

Abstract: The study of the hydrophysical properties of rocks is indispensable for the development of hydraulic engineering, especially for blasting operations in water. Reasonable matching between explosives and rocks increases the utilization of explosive energy and improves the blasting performances. Based on the energy law in the rock blasting process, the matching relationship between explosives and rock is studied by combining experimental and theoretical methods for the hydrophysical properties of the rock itself. Firstly, the theoretical solutions for crushing-zone energy, fragmentation energy and fragment-throwing energy are derived. Subsequently, concrete blocks are prepared with four types of cement–sand ratios, and four types of emulsion explosives are used to carry out single-hole blasting tests in which a high-speed camera is used to capture the trajectory of the blasting fragments that are later collected. Finally, the crushing energy, fracturing energy and fragment-throwing energy are calculated according to the test results and the basic parameters of the used explosives and concrete models. The results show that the size and distribution pattern of blasting blocks are significantly affected by the hydrophysical properties of concrete and explosive properties; the higher the energy consumption in the rupture zone, the smaller the size of the fragments and the more uniform the distribution. Moreover, the median utilization efficiency of explosive energy on rock breaking is 26.4%, the energy consumption in the crushing zone is approximately 8.4%, that in the rupture zone is approximately 10.9%, and that in the throwing energy of fragments accounts for approximately 7.1%. It is also found that the traditional wave impedance matching theory fails to obtain the best explosive energy utilization. On the contrary, the concrete specimen had the best fracturing effect and the highest energy utilization of 30.77% when the impedance ratio of concrete to explosives is 1.479.

Keywords: water; hydrophysical property; explosive; energy match; fragmentation; fracturing zone; crushing zone

check for updates

Citation: Zhu, Z.; Zhou, Z. Experimental Study for the Matching of Explosives and Rocks Based on Rock Hydrophysical Properties. *Water* **2024**, *16*, 1807. https://doi.org/10.3390/w16131807

Academic Editors: Helong Gu, Xueyi Shang and Huatao Zhao

Received: 19 April 2024
Revised: 27 May 2024
Accepted: 4 June 2024
Published: 26 June 2024

1. Introduction

Hydraulic resources are one of the important resources in nature that have a significant impact on human survival activities and engineering operations [1–4]. The correct understanding and reasonable deployment of hydraulic resources are crucial for engineering operations [5–8]. With the development of the social economy, the environment facing the engineering operation is gradually becoming more complex and diversified, and the cognitive requirements for the hydrophysical properties of the working surface are getting higher and higher. Especially for blasting operations, the working environment is sensitive to the effect of relevant environmental characteristicss. Therefore, along with the frequent occurrence of underwater blasting and cave blasting, the matching of explosives and rock has become one of the hot research topics in the field of blasting engineering. A reasonable matching relationship can effectively enhance the energy utilization of explosives, thus improving the blasting effect and reducing the blasting cost. However, the complexity and variability of the rock body, especially in the natural environment, according to the location of the geographic environment and its own water properties reveal the different hydrophysical properties of the rock environment, with different water permeability, softness and water supply. Traditional matching theories, such as wave impedance matching and whole process matching, have relatively limited applications in such complex and variable engineering practices [9,10]. When the rock body is excavated by drilling and blasting methods, part of the energy generated by the explosive blast is used to form the crushing zone, fracture zone and throwing of fragments and causes vibration of plasmas in the elastic zone, and the other part is dissipated in the form of noise, flying rocks and so on. In this process, the water content of the rock has a significant effect. Therefore, it is of great significance to explore the construction of the matching relationship between explosives and rocks from the energy point of view.

Scholars have carried out a series of research studies on the matching problem between explosives and rocks using theoretical, experimental and numerical simulation methods [11–13]. Zhang et al. [14,15] investigated the matching relationship between explosives and rocks using the theory of the equivalent wave resistance method. He proposed that the energy transfer effect could be improved by embedding different impedance media between explosives and rocks. Based on the impedance matching theory, Yang et al. [16] selected high explosive velocity, high-density and high-power explosives for rock breaking in hard rock tunnel drilling and blasting methods. The research results show that the explosive unit consumption and blasting cost are reduced by more than 20%. German scholar Ajay Kumar Jha [17], utilizing the size of the rock impedance to choose a reasonable type of explosive for open-pit limestone mining, achieved better blasting results and productivity. No matter what kind of matching theory is used to match the explosives for different characteristics of the rock body, its ultimate purpose and orientation are to improve the utilization of explosive energy in rock fragmentation. On this understanding, scholars analyzed the relationship between explosives and shock wave impedance matching of rock by reacting to the explosive energy utilization rate through the rock breaking volume. Sanchidrian et al. [18] calculated the percentage of energy used by explosives to produce fragmentation, throw and cause mass vibration by collecting bursting pile, fragment throwing initial velocity and mass vibration data from step blasting excavation tests. The results showed that the fragmentation energy, vibration energy and throwing kinetic energy accounted for 2–6%, 1–3% and 3–21% of the total energy, respectively. In addition, the development of artificial intelligence technology provides more possibilities for explosives and rock matching research. Zhao Mingsheng and Ye Haiwang et al. constructed an explosives and rock matching model based on a neural network algorithm, and the prediction model was successfully applied in blasting engineering [19,20].

In summary, wave impedance matching is the mainstream matching theory. However, with the continuous development of engineering technology, more and more practical projects need to carry out blasting operations under complex and diverse hydraulic and geological conditions, and the site conditions are difficult to support the frequent adjustment of explosive types. Therefore, the limitations of the traditional matching theory gradually appear in these complex backgrounds, and it is difficult to meet the needs of engineering practice for efficient, safe and environmentally friendly blasting. The purpose of this paper is to carry out a series of single-hole blasting concrete model tests from the perspective of energy theory combined with the hydrophysical properties of rocks. In the experiment, using different gray sand ratios formed by the style model, we calculate the energy consumed in the crushing zone, fissure zone, elastic vibration zone and rubble throwing. Thus, the effective utilization rate of explosive energy is obtained, and the mutual influence between different types of explosives, parameters and rock properties is explored in depth so as to construct a reliable matching relationship between explosives and rock for different hydrological environments.

2. Blast Energy Balance Theory

As per the first principle of thermodynamics, the energy released by an explosive blast is carried by the products of the explosion after the chemical reaction is complete. This energy is then converted into heat and work performed on the surrounding medium. It is commonly accepted that the heat and noise generated during the blasting process do not contribute to the dynamic response of the rock and are, therefore, considered useless. The energy involved in the destruction of the rock, on the other hand, is considered useful work. The useful work can be divided into four main parts: (1) The rock is crushed to form a zone with a radius approximately five times that of the hole due to the high burst pressure of explosives around the hole, which is much higher than the dynamic compression strength of the rock; (2) the rock outside the crushing zone is fractured by the dynamic tensile and shear effects, resulting in radial and circumferential fractures, ultimately leading to fragmentation of the rock surface; (3) the rock formation after fracture is influenced by the size and shape of the fragments, which are propelled away from the blast source with kinetic energy; (4) the shock wave created by the blast causes plastic wave propagation within the rock, which ultimately results in elastic wave propagation, manifested as seismic waves or ground surface vibrations. As shown in Figure 1.

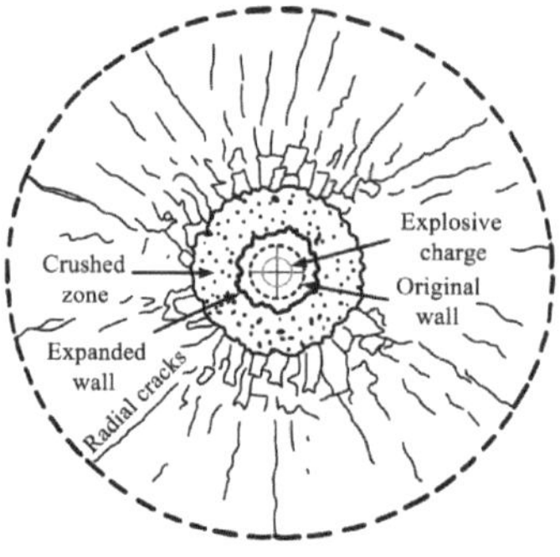

Figure 1. Diagram of the crushed and cracked zones induced by decoupled charge blasting [17].

2.1. Energy in the Crushing Zone

Under the condition of a columnar uncoupled charge structure, the explosive will form a crushed zone with radius R_c around the periphery of the gun hole after the explosion, and the expression for the radius of the crushed zone is as follows:

$$R_c = r_b \left[\frac{AP_d}{\sqrt{2}\sigma_{cd}} \right]^{\frac{1}{a}} \quad (1)$$

$$P_d = \frac{\rho_0 D^2}{8} \left(\frac{r_c}{r_b} \right)^6 n \quad (2)$$

P_{d} is the peak pressure of the borehole wall; r_{b}, r_{c} is divided into the radius of the borehole and charge radius; $A = [(1 + \lambda)^2 + (1 + \lambda^2) - 2\mu(1 - \mu)(1 - \lambda)^2]^{0.5}$, μ for the Poisson's ratio of high-strain loading conditions, $\mu = 0.8\mu_0$, $\lambda = \mu/(1 - \mu)$; σ_{cd} is the medium of the dynamic compression strength; a is the crushed zone shock wave attenuation coefficient, $a = 2 - \mu/(1 - \mu)$; ρ_0 is the rock density; D is the explosive blast velocity; and n is the pressure increase coefficient, generally taken as n = 8~10.

In the crushing area of the rock medium under the action of the shock wave displacement, the diameter of the gun hole increases, and the explosion cavity expands. The shock wave crushes the rock body, and the intensity rapidly decays when it propagates to the boundary of the crushing zone and the entire shock compression process is completed. In this process, the rock complies with the mass conservation criterion, as shown below:

$$(r - r_b^2)\rho_0 = \int_{r_1}^{r} 2\rho r dr \tag{3}$$

r is the radius of the shock wave, r_1 is the radius of the blast cavity corresponding to r, and ρ is the rock density on the surface of the shock wave front.

The shock wave wavefront after the rock density change is negligible, so the air compression of the rock density is ρr instead of ρ at the hole wall, thus obtaining the expansion of the blast cavity law, as follows:

$$r_1 = \left[r^2 - (r^2 - r_b^2)\rho_0/\rho_r\right]^{0.5} \tag{4}$$

$\rho_r = (a + bu)/[a + (b - 1)u]\rho_0$; and a, b is the rock test constant.

When r = R_c, i.e., at the boundary of the crushing zone when the blast expansion process is all over, the final radius of the blast cavity is as follows:

$$R_1 = \left[R_c^2 - (R_c^2 - r_b^2)\rho_0/\rho_r\right]^{0.5} \tag{5}$$

The work performed in the process of the shock wave acting on the crushed zone to produce the burst cavity can be expressed as follows:

$$E_c = \int_{r_b}^{R_1} 2\pi r \sigma_r dr \tag{6}$$

σ_r is the attenuation function of the peak pressure of the shock wave, $\sigma_r = P_d r^{-a}$, and the joint Equation (6) solves the integral function to obtain the expression for the energy in the crushing zone as follows:

$$E_c = 2\pi P_d r_b^2 \left(1 - \frac{r_b}{R_1}\right) \tag{7}$$

2.2. *Rupture Zone Energy*

The shock wave attenuation outside the crushing zone transforms into stress waves that propagate forward in the rock body. This occurs because the dynamic tensile strength of the rock medium is much smaller than its dynamic compression strength. As a result, the stress wave induces new radial cracks and promotes the expansion of cracks generated by the shock wave. It has been demonstrated that the speed of explosive gas propagation is lower than that of stress wave propagation. Therefore, cracks will form in the subsequent gas burst due to penetration, resulting in a large number of fragments.

Assuming that the energy consumption per unit of fracture surface is G_f, the total crushing energy E_f can be calculated according to the following formula:

$$E_f = A_f G_f \tag{8}$$

A_f is the surface area of the fragments produced by blasting. G_f, fracture specific energy, can be calculated in two ways: one is to calculate the Rittinger coefficient by experimental methods, and the other is based on the fracture toughness of the blasted material and elasticity of the theoretical derivation of the model. In single-hole blasting experiments, there are usually hundreds of cracks; so, the use of the inverse of the Rittinger coefficient to characterize the specific energy of fracture is more in line with the actual situation.

The surface area A_f of the fragment can be estimated from the size distribution of the bursting pile. Assuming that the fragment is a sphere of diameter x,

$$A_f = 6V\int_0^{\infty} \frac{f(x)}{x}dx \quad (9)$$

where V is the volume of the fragments and f(x) is the density function of the fragment volume size distribution.

At the end of the blasting experiment, to collect fragments and block size, analysis can be obtained from the particle size distribution curve; according to the volume size, the fragments are grouped, and the introduction of the volume fraction P_k is as follows:

$$p_k = \int_{x_k^I}^{x_k^S} f(x)dx = P(x_k^S) - P(x_k^I) \quad (10)$$

where x_k^I and x_k^S represent the size threshold for class k debris, and P(x) is the cumulative size distribution of the debris.

Bringing Equation (10) into Equation (9), the fragment surface area function can be expressed as follows:

$$A_f = 6V\sum_{k=1}^{C} \int_{x_k^I}^{x_k^S} \frac{f(x)}{x}dx \quad (11)$$

C represents the number of fragment size categories, and *P*(*x*) is the cumulative size distribution of the fragments.

Introducing the logarithmic mean x_k of the size limit of the fragments of class *k*, the transformed form of the fragment surface area function is as follows:

$$A_f = 6V\sum_{k=1}^{C} \frac{p_k}{x_k} \quad (12)$$

2.3. *Kinetic Energy of a Fragment Throw*

Figure 2 shows the blasting fragment by the stress wave and explosive gas coupling to the initial velocity v_0 for oblique throwing motion, the kinetic energy consumed during the flight process with the initial velocity, the end of the velocity, throwing distance and the mass of the fragment. The kinetic energy ΔKE in the horizontal direction of the fragment is as follows:

$$v_2^2 = v_0^2 + 2gL \quad (13)$$

$$\Delta KE = \frac{1}{2}mV_2^2 - \frac{1}{2}mV_1^2 \quad (14)$$

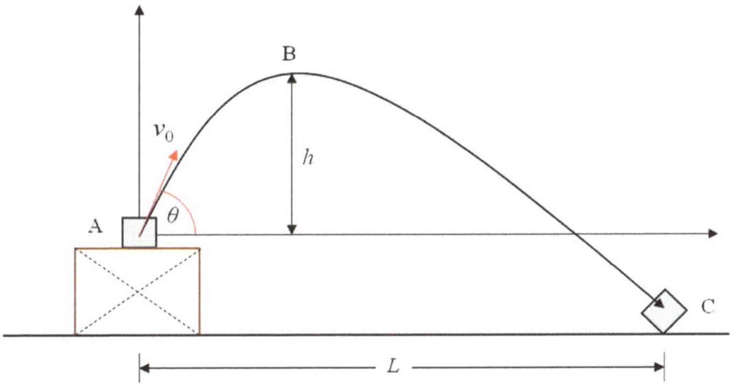

Figure 2. Diagram of the parabola trajectory of fragments.

v_0 is the initial velocity of the fragment; v_2 is the final velocity of the fragment; m is the mass of the fragment; g is the acceleration of gravity; and L is the horizontal throwing distance of the fragment.

The fragment overcomes gravity to rise in the vertical direction to a height of h, so the potential energy can be obtained from Equation (15):

$$\Delta PE = mgh \tag{15}$$

The throws of broken pieces always work for

$$E_t = \Delta KE + \Delta PE \tag{16}$$

3. Single-Hole Blast Modelling Tests

3.1. Pilot Program

In order to study the matching relationship between explosives and rocks, the single-hole blasting model was prepared using cement mortar, and the preparation process is shown in Figure 3a. By changing the ash–sand ratio of the mortar, four kinds of concrete were prepared to simulate four kinds of rocks with different hydrophysical properties and mechanical properties, and the ash–sand and other components were set as shown in Table 1 below. In this way, the effect of the physical and mechanical properties of media with different hydrophysical properties on the crushing zone, fissure zone and throwing energy utilization is investigated.

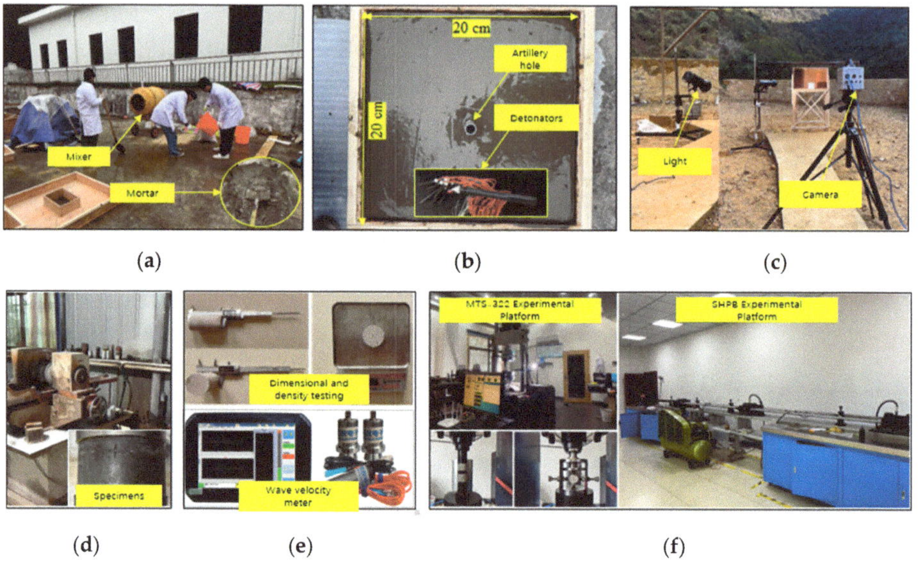

Figure 3. Preparation of blasting models. (**a**) Concrete mixing, (**b**) model casting, (**c**) scene of explosion, (**d**) prototype processing, (**e**) physical parameter testing, (**f**) mechanical test.

Table 1. Components and properties of the concrete specimens.

	Specific Gravity of Each Component			
	Clinker	**Sand**	**Stone**	**Water**
R1	1	2.45	4.12	0.65
R2	1	2.4	3.6	0.65
R3	1	1.95	3.05	0.56
R4	1	1.49	2.54	0.4

For this test, a total of 16 cubic models were prepared, each with side lengths of 20 cm and 30 cm. As depicted in Figure 3b, PVC pipes were used to reserve shell holes with diameters of 12/15 mm and 20/30 cm and lengths. These holes were then filled with expanded ammonium nitrate explosives, which had a charge length of 12 cm and a diameter of 7 mm. The explosives were detonated using instantaneous electronic detonators, as shown in Figure 4. To prevent energy loss caused by the escape of explosive gases, the holes were plugged and compacted with clay at both ends. The plugging length was 4 cm for the 20 cm cube model and 9 cm for the 30 cm cube model. It is important to note that the diameter of the holes in the 30 cm cube model was 15 mm, and the charge diameter and length were the same as those in the 20 cm cube model.

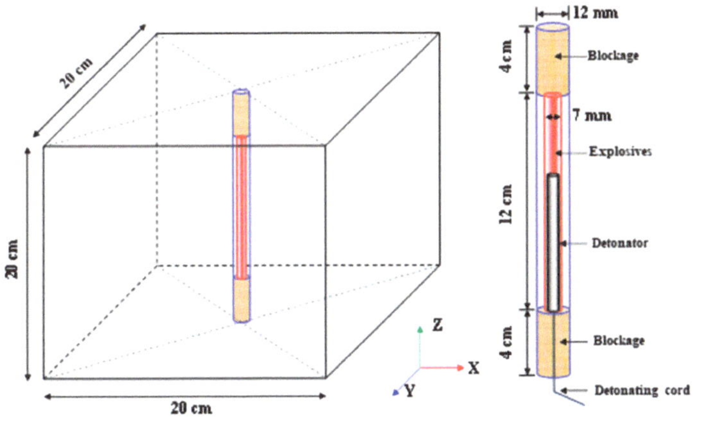

Figure 4. Schematic of blasting model and charge structure.

To eliminate the impact of explosive unit consumption on the test results, we used the same charge density for both model sizes. Additionally, we varied the type of emulsifier and proportion of explosives to obtain four different characteristics of the emulsion explosives, emulsion explosives ratio and physical parameters. These are detailed in Table 2, which shows the matching relationship between explosives and rock. Sixteen single-hole blasting model experiments were conducted using four models of emulsion explosives with varying ash–sand ratios, as detailed in Table 3.

Table 2. Components and properties of the used emulsion explosive.

Number	AN/%	Thiourea/%	Emulsifier/%	Oil Phase/%	ρ_e/kg·m^{-3}	VOD/m·s^{-1}	Q_h/MJ·kg^{-1}	Z/kg·s^{-1}·m^{-2}
E1	78.6	0.2	Span80 (1.8)	Diesel oil 2.4; Engine oil 1.8	1140	3517	3760	4,009,380
E2	78.6	0.2	9126 (1.8)	Diesel oil 2.4; Engine oil 1.8	1170	3643	3870	4,262,310
E3	78.6	0.2	FH17 (1.8)	Diesel oil 2.4; Engine oil 1.8	1160	3936	4060	4,565,760
E4	78.6	0.2	H036 (1.8)	Diesel oil 2.4; Engine oil 1.8	1180	4231	4300	4,992,580

ρ_e is the explosive density; VOD is the detonation velocity; Q_h is the heat of detonation; and Z is the wave impedance.

Table 3. Basic parameters of blasting models and experiment schedule.

Number	V (10^{-3} m^3)	ρ_0 (kg/m^3)	Cp (m/s)	E (GPa)	PF (kg/m^3)	L_c (cm)
R1-E1	7.965	1850.16	3067.65	13.7	1.891	12.0
R1-E2	7.965	1850.16	3067.65	13.7	1.891	12.0
R1-E3	7.965	1850.16	3067.65	13.7	1.891	12.0
R1-E4	7.965	1850.16	3067.65	13.7	1.891	12.0
R2-E1	7.965	1919.27	3284.07	14.4	1.891	12.0
R2-E2	7.965	1919.27	3284.07	14.4	1.891	12.0
R2-E3	7.965	1919.27	3284.07	14.4	1.891	12.0
R2-E4	7.965	1919.27	3284.07	14.4	1.891	12.0
R3-E1	26.947	2086.51	3398.59	16.8	1.891	22.0
R3-E2	26.947	2086.51	3398.59	16.8	1.891	22.0
R3-E3	26.947	2086.51	3398.59	16.8	1.891	22.0
R3-E4	26.947	2086.51	3398.59	16.8	1.891	22.0
R4-E1	26.947	2206.66	3445.74	18.1	1.891	22.0
R4-E2	26.947	2206.66	3445.74	18.1	1.891	22.0
R4-E3	26.947	2206.66	3445.74	18.1	1.891	22.0
R4-E4	26.947	2206.66	3445.74	18.1	1.891	22.0

E1, E2, E3 and E4 in the specimen numbers represent four different ratios of emulsion explosives; R1, R2, R3 and R4 represent four different mortars made with different grey-to-sand ratios; V is the volume of the model; ρ_0 is the density of the concrete after curing; E is the modulus of elasticity of the concrete; PF is the density of the charge; and L_c is the length of the charge.

3.2. Test Results and Analyses

Equations (9)–(12) demonstrate that fragment size and distribution are crucial parameters for solving crushing energy. Therefore, fragments were collected at the end of each blasting test, and the distribution of fragments for the 16 models is shown in Figure 5. The figure reveals that the fragment sizes of the R2-E4, R3-E1 and R1-E2 models are relatively small, with the fragment distribution of the blast pile of R2-E4 being uniform. The R4-E2 and R4-E4 models, however, only break into a few large pieces after an explosion and are accompanied by smaller pieces with a lower percentage of mass.

In order to quantitatively analyze the size of the blasted blocks from the different models, a set of metal screens with apertures arranged in an equal ratio was used to sieve the relatively small-sized fragments, and calipers were used for the larger ones. It should be noted that each set/block of fragments was weighed on an electronic scale, thus obtaining the size and distribution of the fragments after blasting for the different models. The fragment size distribution was fitted using the Swebrec function modified by Ouchterlony [12] with the following functional expression:

$$P(x) = 1 + A[\ln(x_{\max}/x)/\ln(x_{\max}/x_{50})]B + (1-a)\left[\left(\frac{x_{\max}}{x} - 1\right)/\left(\frac{x_{\max}}{x_{50}} - 1\right)\right]^C \quad (17)$$

where x_{max} is the maximum size at $P(x_{max})$ = 100%, which can be obtained by directly measuring the size of the largest fragment; x_{max} is the median size at $P(x_{50})$ = 50%; A is the grading coefficient, with a range of values from 0 to 1; and B and C are fitting parameters.

After the collection of the blast pile, the relatively small-sized fragments were sieved using a set of metal screens with apertures arranged in an equiproportional manner, and the larger sizes were measured using calipers. It is important to note that each set/block of debris was placed on an electronic scale for weighing, thus obtaining the particle size distribution of the debris after blasting for the different models, as shown in Figure 6.

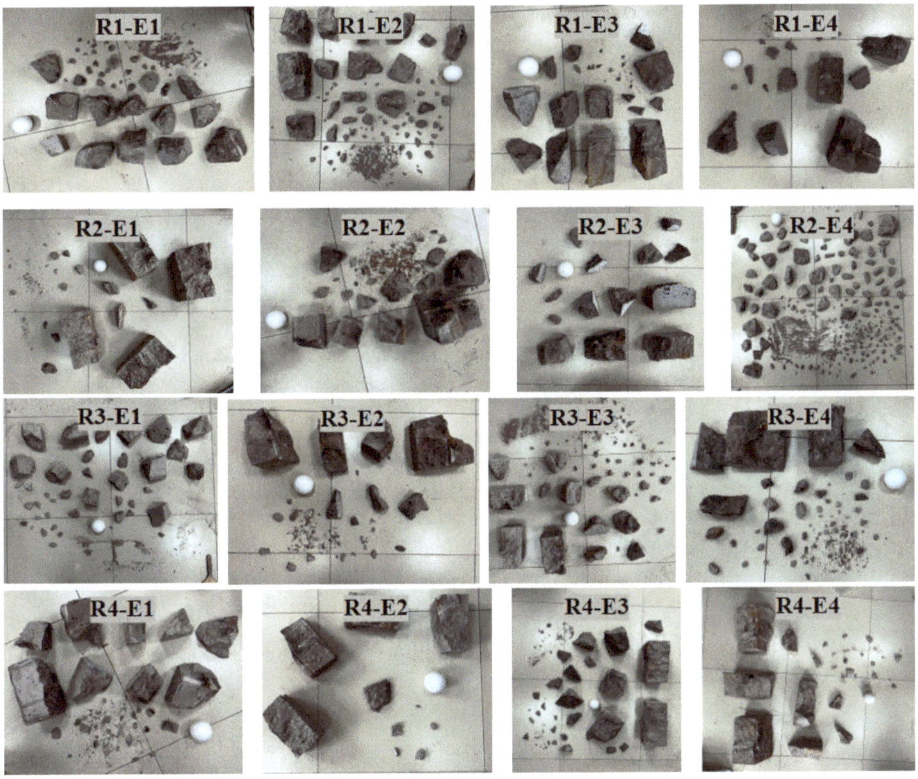

Figure 5. Fragments induced by single-hole blasting.

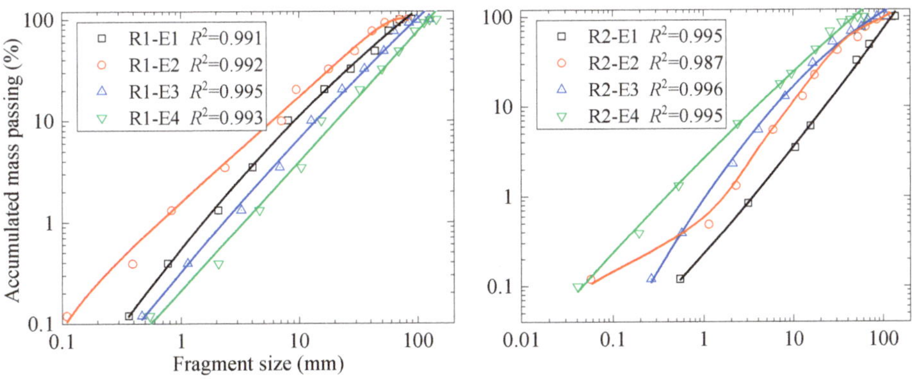

Figure 6. Curves of accumulated mass passing vs. fragment size.

Statistics on the block sizes corresponding to the different models at a percentage of 1%, 10% and 50% are shown in Table 4.

Table 4. Results of post-blast fragmentation.

Number	x_1 (mm)	x_{10} (mm)	x_{50} (mm)	Model Number	x_1 (mm)	x_{10} (mm)	x_{50} (mm)
R1-E1	1.42	7.89	45.64	R3-E1	1.62	7.54	35.64
R1-E2	0.67	7.01	29.20	R3-E2	4.67	12.01	46.20
R1-E3	2.74	12.53	53.41	R3-E3	2.34	10.53	43.14
R1-E4	3.87	15.02	72.32	R3-E4	5.87	12.66	52.32
R2-E1	3.34	18.42	75.47	R4-E1	3.34	8.96	35.47
R2-E2	0.65	7.77	34.56	R4-E2	5.45	9.77	54.56
R2-E3	3.02	6.98	27.12	R4-E3	1.22	6.48	22.18
R2-E4	0.43	5.61	19.43	R4-E4	4.43	18.61	79.65

x_1, x_{10}, x_{50} denote particle sizes with a cumulative percentage of 1 percent, 10 percent and 50 percent.

From Equations (13)–(16), it can be observed that, in order to solve the blast throwing energy, the mass, particle size, horizontal and vertical throwing distance and initial flight speed of the thrown fragments must be obtained. Traditional manual statistical methods cannot capture the flight speed and vertical throwing height of the debris; so, we used high-speed cameras during the blasting process to record the flight video of the debris, and then used Tracker video analysis software to identify the debris and track the trajectory of the debris so as to obtain the debris in the throwing process of the initial speed of the vertical and horizontal throwing distance parameters.

The explosive parameters in Table 2 are brought into Equation (17), and the total explosive energy E_0 of each model can be obtained by combining the loading structure of the single-hole model, which is listed in Table 5. At the same time, the joint crushing zone and fissure zone energy solution function can be obtained in each model after blasting the crushing energy E_c, rupture energy E_f, throwing energy E_t and its proportion of each part in the total energy of the explosion, as shown in Table 5.

Table 5. Energy consumption of each component and efficiencies.

Number	E_0 (KJ)	E_c (KJ)	η_c	E_f (KJ)	η_f	E_t (KJ)	η_t	η
R1-E1	56.64	4.39	7.75%	6.55	11.6%	4.045	7.1%	26.46%
R1-E2	58.30	5.12	8.78%	7.33	12.6%	4.665	8.0%	29.36%
R1-E3	61.16	4.07	6.65%	6.82	11.2%	4.021	6.6%	24.38%
R1-E4	64.78	4.13	6.38%	6.55	10.1%	3.875	6.0%	22.47%
R2-E1	56.64	4.08	7.20%	5.25	9.3%	3.335	5.9%	22.36%
R2-E2	58.30	4.66	7.99%	6.06	10.4%	3.745	6.4%	24.81%
R2-E3	61.16	5.72	9.35%	6.85	11.2%	4.425	7.2%	27.79%
R2-E4	64.78	6.27	9.68%	8.90	13.7%	4.765	7.4%	30.77%
R3-E1	191.64	18.61	9.71%	24.08	12.6%	14.76	7.7%	29.98%
R3-E2	197.25	15.91	8.07%	19.38	9.8%	13.53	6.9%	24.75%
R3-E3	206.93	21.58	10.43%	24.32	11.8%	15.55	7.5%	29.70%
R3-E4	219.16	18.81	8.58%	20.52	9.4%	13.46	6.1%	24.09%
R4-E1	191.64	17.57	9.17%	19.84	10.4%	15.85	8.3%	27.79%
R4-E2	197.25	16.28	8.25%	18.41	9.3%	15.19	7.7%	25.29%
R4-E3	206.93	19.79	9.56%	23.22	11.2%	16.47	8.0%	28.74%
R4-E4	219.16	16.12	7.36%	22.2	10.1%	14.01	6.4%	23.88%

E_e is the chemical energy of the explosive; η_c is the share of crushing energy; η_f is the share of fracturing energy; η_t is the share of throwing energy; and η is the effective utilization rate of the energy used in blasting to break up the rock.

As can be observed in Table 5, the average value of the total energy generated by the explosives of the cube model with a side length of 20 cm is 60.2 KJ, the total energy of the explosion of the E4 explosives is 8.2 KJ more than that of the E1 explosives for the charge developed in this type of model, the average value of the total energy of the explosion of the explosives of the cube model with a side length of 30 cm is 203.75 KJ, and the total energy of the explosives of the E4 explosives generated by the E4 explosives is 27.5 KJ more than that of the E1 explosives under the specific charging conditions.

From Figure 7, can be learned from the different explosives in the same kind of concrete used to crush, rupture and broken pieces thrown in each part of the energy proportion of the change in the law. For R1-type hydrophysical properties of concrete (the lowest mechanical strength), the effective utilization rate of explosive energy η with the enhancement of explosive properties (density and wave velocity are increasing) show a trend of increasing and then decreasing. The change rule of energy utilization of each component is basically the same. It can be observed that R1-type concrete matched with E2-type explosives has the highest energy utilization rate of 29.36%, the rupture energy accounted for 12.57%, and the concrete and explosives wave impedance ratio was 1.243. For R2-type hydrophysical properties of the concrete, explosive energy utilization with the enhancement of explosive characteristics increases gradually; the maximum value of explosive effective utilization occurs in the case of matching with the E4-type explosives, at which point η = 30.77% and the ratio of wave impedance of concrete and explosives is 1.479. The energy utilization of crushed zone ηc = 7.2%, and the total energy utilization η = 27.79% when using E1 explosives for blasting and excavation of R2 concrete.

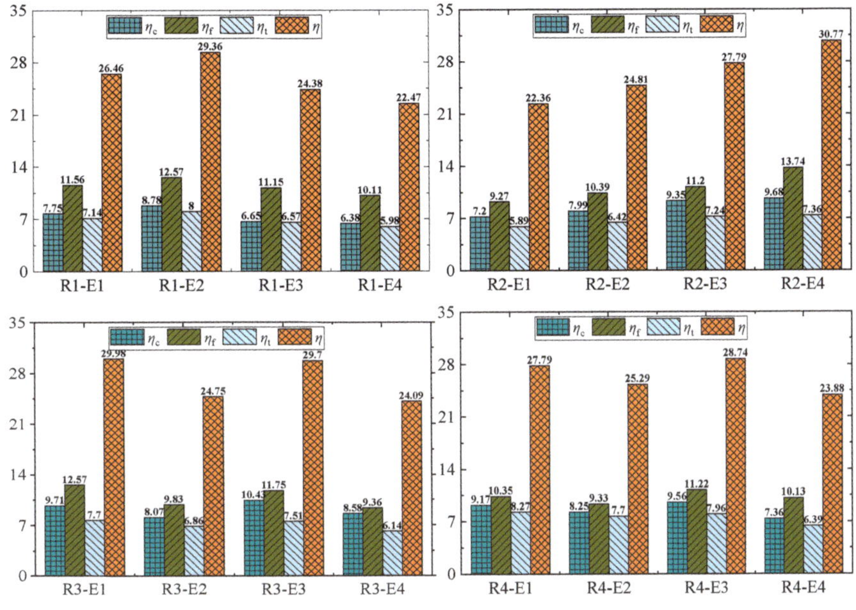

Figure 7. Distributions of blasting energy.

For R3 concrete with hydrophysical properties, the effective energy utilization of explosives for blasting excavation using E1 and E3 explosives was essentially the same, approximately 30%. The fracturing energy Ec in the R3-E1 matched explosion model accounts for 32.4% of the effectively utilized energy. At this time, the wave impedance ratio is 1.523. For R4 hydrophysical properties of concrete, the use of E3 explosives can achieve the maximum effective utilization of energy, 28.74%. At this time, the proportions of crushing energy, rupture energy and throwing energy are 9.56%, 11.22% and 7.96%, respectively, and the value of the wave impedance ratio is 1.784. The above conclusions show that both the explosives and the exploded medium have a very obvious effect on the utilization of explosive energy. In this research test, the highest explosive energy utilization of 30.77% was obtained when R2-type concrete and E4-type explosives were matched. In addition, the total energy consumed in the process of crack expansion to produce fragments was always greater than the energy consumed in the crushing zone. This is in general agreement with the findings of Sanchidrian et al. [18].

4. Conclusions

1. We assume that the energy consumed in the crushing zone, rupture zone and fragment throwing is the effective energy of the work performed on the rock by explosive blasting. Based on the mass conservation criterion in the crushing zone, the total rupture energy is defined by the product of the surface fracture energy and the fracture surface area. In consideration of the kinetic energy consumed in the process of throwing fragments and overcome the gravitational potential energy to do work, the theoretical derivation of the functional expression to solve the energy of each part.
2. Different explosives and different hydrophysical conditions will lead to different size and distribution of fragments after detonation. The average values of x1, x10 and x50 produced by the single-hole blasting model are 2.81 mm, 10.49 mm and 45.39 mm, and the best blasting and crushing effect is obtained when the hydrophysical properties of concrete and E4 emulsion explosives are matched with R2 type, and the minimum size of the block is 0.041 mm and the maximum size is 62.1 mm, which provides a certain reference for the study of blasting explosives and rock-matching system of water conservancy projects. For the water conservancy project blasting explosives rock matching system research provides a certain reference.

Author Contributions: Conceptualization, Z.Z. (Zhou Zhiyong); Formal analysis, Z.Z. (Zhu Zhaozhen); Investigation, Z.Z. (Zhu Zhaozhen); Writing—original draft, Z.Z. (Zhu Zhaozhen); Writing—review & editing, Z.Z. (Zhu Zhaozhen). All authors have read and agreed to the published version of the manuscript.

Funding: This research received no external funding.

Data Availability Statement: Data available in a publicly accessible repository that does not issue DOIs.

Conflicts of Interest: The authors declare no conflict of interest.

References

1. Gaagai, A.; Aouissi, H.A.; Bencedira, S.; Hinge, G.; Athamena, A.; Heddam, S.; Gad, M.; Elsherbiny, O.; Elsayed, S.; Eid, M.H.; et al. Application of Water Quality Indices, Machine Learning Approaches, and GIS to Identify Groundwater Quality for Irrigation Purposes: A Case Study of Sahara Aquifer, Doucen Plain, Algeria. *Water* **2023**, *15*, 289. [CrossRef]
2. Gad, M.; Gaagai, A.; Eid, M.H.; Szűcs, P.; Hussein, H.; Elsherbiny, O.; Elsayed, S.; Khalifa, M.M.; Moghanm, F.S.; Moustapha, M.E.; et al. Groundwater Quality and Health Risk Assessment Using Indexing Approaches, Multivariate Statistical Analysis, Artificial Neural Networks, and GIS Techniques in El Kharga Oasis, Egypt. *Water* **2023**, *15*, 1216. [CrossRef]
3. Gu, H.; Lai, X.; Tao, M.; Cao, W.; Yang, Z. The role of porosity in the dynamic disturbance resistance of water-saturated coal. *Int. J. Rock Mech. Min. Sci.* **2023**, *166*, 105388. [CrossRef]
4. Gu, H.; Lai, X.; Tao, M.; Momeni, A.; Zhang, Q. Dynamic mechanical mechanism and optimization approach of roadway surrounding coal water infusion for dynamic disaster prevention. *Measurement* **2023**, *223*, 113639. [CrossRef]
5. Guo, Z.T.; Wu, C.S. Matching of explosives with rock in the whole process. *Min. Metall. Eng.* **1993**, *13*, 11–15.
6. Haider, S.; Masood, M.U.; Rashid, M.; Alshehri, F.; Pande, C.B.; Katipoğlu, O.M.; Costache, R. Simulation of the Potential Impacts of Projected Climate and Land Use Change on Runoff under CMIP6 Scenarios. *Water* **2023**, *15*, 3421. [CrossRef]
7. Hong, Z.; Tao, M.; Cui, X.; Wu, C.; Zhao, M. Experimental and numerical studies of the blast-induced overbreak and underbreak in underground roadways. *Undergr. Space* **2023**, *8*, 61–79. [CrossRef]
8. Hu, Y.; Lu, W.; Chen, M.; Yan, P.; Yang, J. Comparison of Blast-Induced Damage Between Presplit and Smooth Blasting of High Rock Slope. *Rock Mech. Rock Eng.* **2014**, *47*, 1307–1320. [CrossRef]
9. Ilyushin, Y.V.; Asadulagi, M.A.M. Development of a Distributed Control System for the Hydrodynamic Processes of Aquifers, Taking into Account Stochastic Disturbing Factors. *Water* **2023**, *15*, 770. [CrossRef]
10. Liu, M.X.; Zhang, Y.P.; Nie, X.J. Experimental study on the matching between explosive properties and rock. *Eng. Blasting* **2016**, *22*, 24–29.
11. Mancosu, N.; Snyder, R.L.; Kyriakakis, G.; Spano, D. Water Scarcity and Future Challenges for Food Production. *Water* **2015**, *7*, 975–992. [CrossRef]
12. Ouchterlony, F. The Swebrec((c)) function: Linking fragmentation by blasting and crushing. *Min. Technol.* **2005**, *114*, 29–44. [CrossRef]
13. Pershin, I.M.; Papush, E.G.; Kukharova, T.V.; Utkin, V.A. Modeling of Distributed Control System for Network of Mineral Water Wells. *Water* **2023**, *15*, 228. [CrossRef]

14. Zhang, Z.X. Kinetic energy and its applications in mining engineering. *Int. J. Min. Sci. Technol.* **2017**, *27*, 237–244. [CrossRef]
15. Zhang, Z.-X.; Ouchterlony, F. Energy Requirement for Rock Breakage in Laboratory Experiments and Engineering Operations: A Review. *Rock Mech. Rock Eng.* **2022**, *55*, 629–667. [CrossRef]
16. Yang, N.H.; Zhang, Z.Y. Drilling and blasting technology for rapid tunnel boring in hard rock tunnels. *Eng. Blasting* **2003**, *9*, 16–21.
17. Jha, A.K. Impedance Matching Algorithm for Selection of Suitable Explosives for Any Rock Mass—A Case Study. *J. Geol. Resour. Eng.* **2020**, *8*, 11.
18. Sanchidrián, J.A.; Segarra, P.; López, L.M. Energy components in rock blasting. *Int. J. Rock Mech. Min. Sci.* **2007**, *44*, 130–147. [CrossRef]
19. Ye, H.W. Optimization system of explosives and rock matching based on fuzzy neural network. *Blasting Equip.* **2005**, *11*, 5–7.
20. Zhao, M.S.; Xu, H.B. Research on the matching of explosives and rocks based on neural network. *J. Liaoning Inst. Sci. Technol.* **2009**, *11*, 1–3.

water

MDPI

Article

Coal-Mine Water-Hazard Risk Evaluation Based on the Combination of Extension Theory, Game Theory, and Dempster–Shafer Evidence Theory

Xing Xu [1,2], Xingzhi Wang [1,*] and Guangzhong Sun [2]

[1] School of Geoscience and Technology, Southwest Petroleum University, Chengdu 610500, China
[2] School of Resources and Safety Engineering, Henan University of Engineering, Zhengzhou 451191, China
* Correspondence: wxzswpi@163.com

Abstract: Due to the complex hydrogeological conditions and water hazards in coal mines, there are multiple indexes, complexities, incompatibilities, and uncertainty issues in the risk evaluation process of coal-mine water hazards. To accurately evaluate the risk of coal-mine water hazards, a comprehensive evaluation method based on extension theory, game theory, and Dempster–Shafer (DS) evidence theory is proposed. Firstly, a hierarchical water-hazard risk-evaluation index system is established, and then matter-element theory in extension theory is used to establish a matter-element model for coal-mine water-hazard risk. The membership relationship between various evaluation indexes and risk grades of coal-mine water-hazard risk is quantified using correlation functions of extension set theory, and the quantitative results are normalized to obtain basic belief assignments (BBAs) of risk grades for each index. Then, the subjective weights of evaluation indexes are calculated using the order relation analysis (G1) method, and the objective weights of evaluation indexes are calculated using the entropy weight (EW) method. The improved combination weighting method of game theory (ICWMGT) is introduced to determine the combination weight of each evaluation index, which is used to correct the BBAs of risk grades for each index. Finally, the fusion of DS evidence theory based on matrix analysis is used to fuse BBAs, and the rating with the highest belief fusion result is taken as the final evaluation result. The evaluation model was applied to the water-hazard risk evaluation of Sangbei Coal Mine, the evaluation result was of II grade water-hazard risk, and it was in line with the actual engineering situation. The evaluation result was compared with the evaluation results of three methods, namely the expert scoring method, the fuzzy comprehensive evaluation method, and the extension method. The scientificity and reliability of the method adopted in this paper were verified through this method. At the same time, based on the evaluation results, in-depth data mining was conducted on the risk indexes of coal-mine water hazards, and it was mainly found that 11 secondary indexes are the focus of coal-mine water-hazard risk prevention and control, among which seven indexes are the primary starting point for coal-mine water-hazard risk prevention and control. The groundwater index in particular has the most prominent impact. These results can provide a theoretical basis and scientific guidance for the specific water-hazard prevention and control work of coal mines.

Keywords: coal-mine water hazard; extension theory; improved combination weighting method of game theory; Dempster–Shafer evidence theory; risk evaluation

Citation: Xu, X.; Wang, X.; Sun, G. Coal-Mine Water-Hazard Risk Evaluation Based on the Combination of Extension Theory, Game Theory, and Dempster–Shafer Evidence Theory. *Water* **2024**, *16*, 2881. https://doi.org/10.3390/w16202881

Academic Editor: Chin H Wu

Received: 5 September 2024
Revised: 22 September 2024
Accepted: 9 October 2024
Published: 10 October 2024

1. Introduction

Chinese coal production accounts for about 50% of the world's total, and its raw coal production has been ranked first in the world for many consecutive years. The distribution of coal in China is extensive, and China's land is composed of multiple tectonic plates that have undergone multiple geological tectonic movements. As a result, the water filling conditions for coal mines in China are extremely complex [1,2]. In recent years, with the extension of mining levels and the expansion of mining scope, the mining of coal seams has faced the presence of water accumulation in the goaf above the roof and high-pressure water near the bottom plate, which increases the control factors for water inrush. The mechanism and types of water inrush are also complex and varied, and the threat of water hazards to mines has become increasingly severe [3,4]. Mine water hazards seriously threaten the safety production of coal mines and the life safety of underground workers, which has caused great losses to workers and coal mines [5]. According to incomplete statistics, there were a total of 1103 water accidents in China from 2001 to 2022; these accidents resulted in 4667 deaths [6], which seriously threatens the efficient development process of China's coal resources. How to prevent and control coal-mine water hazards is one of the key issues that urgently needs to be addressed. The sixteen-character principle of "prediction and forecasting, investigation if suspected, exploration before excavation, and treatment before mining" and the comprehensive water control measures of "prevention, blocking, dredging, drainage, and interception" have been proposed for the prevention and control of water hazards in coal mines [7]. It can be seen that prediction and forecasting are the primary link and foundation of coal-mine water-hazard prevention and control. The scientific and reasonable prediction and forecasting of coal-mine water hazards, as well as the evaluation of the risk grade of coal-mine water hazards, have practical guiding significance for guiding water-hazard prevention and control work in coal mines.

The occurrence of coal-mine water accidents is a result of the comprehensive influence and control of multiple factors based on hydrogeological conditions. The factors causing coal-mine water hazards can generally be divided into two categories: water filling sources and water diversion channels. Different combinations of water filling sources and water diversion channels will constitute different mine water-filling intensities and water-hazard threat grades [8–10]. Based on the main causes of water hazards, coal technology workers have conducted extensive research on the prediction, forecasting, and risk evaluation of coal-mine water hazards from different perspectives and directions. The impact of mine water sources on the occurrence of mine water inrush is reflected in their provision of water sources for mining space, which is the fundamental prerequisite for the occurrence of water inrush. Water filling sources mainly include atmospheric precipitation, surface water, groundwater, and old goaf water, and the methods for identifying water filling sources mainly include geological and hydrogeological analysis, water temperature and level analysis, hydrochemical analysis, mathematical theory analysis, etc. [11]. The hydrochemical characteristics of groundwater are strictly controlled by factors such as recharge sources, transport conditions, and the lithological characteristics of the storage space. Therefore, groundwater chemistry can reflect the most essential characteristics of groundwater. If the hydrochemical characteristics of a target aquifer that may become a water filling source are analyzed, the risk grade of water filling in the mine can be accurately evaluated, and water-hazard prevention measures can be proposed based on it [12–15]. Mine water-diversion channels are another prerequisite for the occurrence of mine water accidents. As an important channel for communicating the water filling source and mining space, it is an important factor affecting the evolution of the original aquifer structure under mining conditions. The water diversion channels in mines can be roughly divided into two types: natural channels and artificial channels [10]. Natural channels include fault zones, water-conducting collapse columns, hidden outcrops, etc. Based on the analysis of a large number of water inrush accidents, more than 80% of them are related to the faults [16,17]. Many engineering examples also show that the existence of faults reduces the strength of rocks, and the vast majority of non-water-conducting faults are activated during coal seam mining, which results in the continuous expansion of fractures and provides conditions for the formation of water diversion channels. Studying the activation law of faults provides a theoretical basis for predicting and forecasting mine water inrush risk and scientific management [3,18–21]. Artificial channels include mining induced fractures, ground karst collapse zones, and poorly sealed boreholes. During the process of coal seam mining, rock layers around the goaf lose their support, which causes changes in the original stress state and results in the destruction of the roof, floor, and coal walls. This forms mining-induced fractures around the mining area, and if mining-induced fractures develop into aquifers, they can cause water inrush in mines. Studying the development law of mining-induced fractures has important guiding significance for predicting and forecasting mine water-hazard risk [22–26].

The above is mainly based on one of the two major types of disaster-causing factors for mine water-hazard prediction and risk evaluation. Some scholars have also combined different theories and methods based on the two types of disaster-causing factors to predict and evaluate water hazards locally or globally in mines. Bai et al. [5] used analytic hierarchy process and criteria importance though intercriteria correlation to determine the comprehensive weights of seven factors; they established a risk index model for mine water inrush and identified a risk area of roof water inrush in the No.1 coal seam of Liuzhuang Coal Mine. Chen et al. [27] selected six main influencing factors as discriminant indexes and established a Fisher discriminant model based on multiple training samples, which accurately predicted water inrush risk areas of three main coal seams in the southern mining area of Qidong Coal Mine. Huang et al. [28] used analytic hierarchy process and the entropy weight method to calculate the weights of six indexes; they established a vulnerability index model for bottom-plate water inrush and successfully predicted the potential water risk area of the 182,602 working face. Ruan et al. [29] used an improved analytic hierarchy process method to calculate the weights of thirty evaluation indexes, and a decision model for underground water inrush in mines was established combining DS evidence theory. They also predicted the probability of water inrush in the 20,101 ventilation roadway of Wangjialing Coal Mine. Wang et al. [30] conducted a special evaluation of water inrush risk in six coal mines around Yangzhuang Coal Mine using the safety checklist method. Wang et al. [31] evaluated the risk of mine water inrush accidents using analytic hierarchy process and an expert scoring method. Zhao [32] evaluated the degree of water damage threat to a mine in Ningdong Mining Area using the analytic hierarchy process method and fuzzy comprehensive evaluation method. Xu et al. [33] evaluated the water-hazard risk of Zhaojiazhai Coal Mine based on the combination weighting method and fuzzy comprehensive evaluation method; they proposed suggestions for water-hazard prevention and control. Li et al. [34] used an evaluation model combining the analytic network hierarchy process method and a cloud model to conduct a water inrush risk warning evaluation of Y Coal Mine, and the evaluation model is now available and reliable. Chen et al. [35] constructed an evaluation model for coal-mine water inrush risk based on the fuzzy network analysis method.

Conducting mine water-hazard risk evaluation and understanding the water-hazard risk grade of mines in a timely manner can provide decision-making support for coal-mine water-hazard prevention and control work and promote the improvement of management level and efficiency of coal mines. It also serves as a favorable tool for higher-level supervisory departments to grasp the safety production status of coal mines and strengthen safety supervision. Although many scholars have used various evaluation methods to analyze mine water-hazard risk, which have played an important guiding role in preventing and reducing water accidents, there are more or less the following problems: firstly, the selection of evaluation indexes is not representative, and the establishment of evaluation index system is not scientific and perfect, which may introduce certain difficulties to the promotion of this method. Secondly, the weight calculation method of evaluation indexes is unreasonable, and the imbalance of weight calculation is caused by a single weighting method, which leads to a great increase in uncertainty in the evaluation process. Thirdly, there is a lack of scientific rigor in the selection of comprehensive evaluation methods, as well as a lack of consideration for inherent connections between evaluation indexes—these may also lead to deviations between evaluation results and engineering reality. Therefore, it is particularly necessary to establish a scientific and reasonable risk-evaluation method for mine water hazards. Based on the above reasons, this paper proposes a coal-mine water-hazard risk-evaluation method based on the combination of extension theory, game theory and DS evidence theory, in order to accurately evaluate water-hazard risk.

2. Basic Theories and Evaluation Methods

2.1. Extension Theory

Extension theory is a theoretical framework based on matter-element theory and extension set theory. It cannot only use the extensibility of matter elements to determine qualitative methods, but it also uses extension set theory to perform quantitative calculations through correlation functions. Based on the extensibility of matter elements, many problems described as contradictions can be solved under the matter-element model [36–38]. The specific analysis steps are as follows:

2.1.1. Establishing Matter-Element Model

Firstly, the evaluation matter-element R is determined, where R is the ordered ternary group composed of "matter, characteristic, value", $R = \begin{pmatrix} N & C & V \end{pmatrix}$. Among them, N represents the matter of matter elements, C represents the characteristic of matter elements, and V represents the value of matter elements; it is a basic element matter to describe an object. Then, the classical domain matter-element R_j, node domain matter-element R_p and matter-element to be evaluated R_0 are determined, respectively.

$$R_j = \begin{pmatrix} N_j & c_i & v_{ji} \end{pmatrix} = \begin{bmatrix} N_j & c_1 & v_{j1} \\ & c_2 & v_{j2} \\ & \vdots & \vdots \\ & c_n & v_{jn} \end{bmatrix} = \begin{bmatrix} N_j & c_1 & [a_{j1}, b_{j1}] \\ & c_2 & [a_{j2}, b_{j2}] \\ & \vdots & \vdots \\ & c_n & [a_{jn}, b_{jn}] \end{bmatrix} \tag{1}$$

where R_j represents the classical domain matter-element model of the jth evaluation grade of evaluation object, $(j = 1, 2, \cdots, m)$. N_j represents the jth evaluation level of the evaluation object; c_i represents the ith evaluation index $(i = 1, 2, \cdots, n)$; and $v_{ji} = [a_{ji}, b_{ji}]$ represents the range of the value taken by the jth evaluation level with respect to the ith evaluation index.

$$R_p = \begin{pmatrix} N_p & c_i & v_{pi} \end{pmatrix} = \begin{bmatrix} N_p & c_1 & v_{p1} \\ & c_2 & v_{p2} \\ & \vdots & \vdots \\ & c_n & v_{pn} \end{bmatrix} = \begin{bmatrix} N_j & c_1 & [a_{p1}, b_{p1}] \\ & c_2 & [a_{p2}, b_{p2}] \\ & \vdots & \vdots \\ & c_n & [a_{pn}, b_{pn}] \end{bmatrix} \tag{2}$$

where R_p represents a characteristic range of the joint domain matter-element model for all evaluation grades of the evaluation object. N_p represents all the evaluation grades of the evaluation object. v_{pi} represents the value range of evaluation index c_i with respect to all grades, $v_{pi} = [a_{pi}, b_{pi}]$.

$$R_0 = \begin{pmatrix} N_0 & c_i & v_i \end{pmatrix} = \begin{bmatrix} N_0 & c_1 & v_1 \\ & c_2 & v_2 \\ & \vdots & \vdots \\ & c_n & v_n \end{bmatrix} \tag{3}$$

where R_0 represents the matter-element model determined according to the evaluation object. N_0 represents the evaluation object, and v_i represents that evaluation object. N_0 corresponds to the assignment of evaluation index c_i.

2.1.2. Calculating Correlation Functions

After determining the above three matter-element models, in order to express the variation range of an index with respect to each evaluation grade—that is, to describe the quantitative and qualitative change of matter elements—the concept of a correlation function is introduced, and its calculation equation is defined as

$$K_j(v_i) = \begin{cases} \frac{-\rho(v_i, v_{ji})}{|v_{ji}|} & v_i \in v_{ji} \\ \frac{\rho(v_i, v_{ji})}{\rho(v_i, v_{pi}) - \rho(v_i, v_{ji})} & v_i \notin v_{ji} \end{cases} \tag{4}$$

where $\rho(v_i, v_{ji}) = \left|v_i - \frac{a_{ji}+b_{ji}}{2}\right| - \frac{b_{ji}-a_{ji}}{2}$, $\rho(v_i, v_{pi}) = \left|v_i - \frac{a_{pi}+b_{pi}}{2}\right| - \frac{b_{pi}-a_{pi}}{2}$, $K_j(v_i)$ represents the correlation degree between the ith index and the jth grade of evaluation object N_0, $\rho(v_i, v_{ji})$ represents the distance between the evaluation index value of matter-element R_0 and the classical domain matter-element R_j, and $\rho(v_i, v_{pi})$ represents the distance between the evaluation index value of matter-element R_0 and the joint domain matter-element R_p.

2.2. Improved Combination Weighting Method of Game Theory

At present, the main weighting methods are divided into the subjective weighting method, the objective weighting method and the combination weighting method [5,39]. The subjective weighting method mainly depends on the experience of experts and cannot well reflect objective reality; the objective weighting method has strong objectivity and calculation accuracy, but it ignores the positive subjective initiative of experts. The combination weighting method can overcome the shortcomings of the single weighting method and obtain the optimal weight combination, taking into account the characteristics of the different weighting methods. It can improve the scientific rationality of the evaluation. The subjective and objective combination weighting method based on improved game theory is used to determine combination weights of evaluation indexes, where the subjective weighting method chooses the G1 method and the objective weighting method chooses the EW method.

2.2.1. Order Relation Analysis Method

The G1 method is a subjective weighting method proposed by Chinese scholar Professor Guo Yajun [40]. This method determines the subjective weight of the index according to the importance relationship between the adjacent evaluation indexes. Compared with the traditional analytic hierarchy process, its evaluation process is clear and does not need to build a judgment matrix and consistency test, which avoids the large amount of calculation and tedious calculation process. The calculation steps are as follows:

(1) Determining the order relation.

Relative to a certain evaluation object, the set of each evaluation index is $\{c_i\}$, and c_i is the ith evaluation index, ($i = 1, 2, \cdots, n$). According to the subjective experience, the expert selects the largest value in evaluation index set, which is marked as c_1^*, and the second-most-important evaluation index is selected among the remaining $n-1$ evaluation indexes, which is recorded as c_2^*. If the last evaluation index is recorded as c_n^*, the order relation of evaluation indexes is established as

$$c_1^* \geq c_2^* \geq \cdots \geq c_n^* \tag{5}$$

(2) Giving the ratio of the relative importance.

If the ratio of the importance degree of evaluation indexes c_{k-1}^* and c_k^* in the order relation is $\frac{w_{k-1}^{(1)*}}{w_k^{(1)*}}$, $w_k^{(1)*}$ is the weight of the kth evaluation index of the subjective weighting method, the weight ratio of the adjacent evaluation index can be expressed as

$$\frac{w_{k-1}^{(1)*}}{w_k^{(1)*}} = r_k(k = n, n-1, \cdots, 2) \tag{6}$$

where r_k is the assignment of the importance of the order relation regarding evaluation indexes c_{k-1}^* and c_k^*, and the r_k assignment is shown in Table 1.

Table 1. Assignment table.

r_k	Importance of Evaluation Index c_{k-1}^* to c_k^*
1	Equally Important
1.2	Slightly Important
1.4	Obviously Important
1.6	Strongly Important
1.8	Extremely Important
1.1, 1.3, 1.5, 1.7	The median of the above two adjacent judgments

(3) Determining the subjective weights.

The weight equation of evaluation index c_n^* is defined as

$$w_n^{(1)*} = \left(1 + \sum_{k=2}^{n} \prod_{i=k}^{n} r_k\right)^{-1} \tag{7}$$

According to the recursive equation, the weights of other indexes can be determined as

$$w_{k-1}^{(1)*} = r_k w_k^{(1)*} \tag{8}$$

Because $w_k^{(1)*}$ is the weight corresponding to the order relation, it needs to adjust the weight according to the corresponding order of the original evaluation indexes $c_1, c_2, \cdots, c_n$, and the adjustment result is $w_k^{(1)}$.

2.2.2. Entropy Weight Method

The EW method is an objective weighting method [41]. Its core idea is to determine the weight of each index datapoint based on the discretization degree of each index datapoint. If the discretization degree of index data is greater, it means that the index contains more information and has a great influence on the decision-making results. Therefore, it will be given a greater weight, and vice versa. The calculation steps are as follows:

(1) Constructing an evaluation index matrix.

There are m evaluation objects and n evaluation indexes, and the evaluation index data are b_{ji}, $(j = 1, 2, \cdots, m; i = 1, 2, \cdots, n)$. The index matrix is defined as

$$B = \left[b_{ji}\right]_{m \times n} \tag{9}$$

Matrix B is standardized by Equations (10) and (11), and the standardized matrix B' is obtained, $B' = \left[b'_{ji}\right]_{m \times n}$, b'_{ji} is the standardization result of b_{ji}.

For the index of the bigger the better (income-type attribute), the standardized equation is defined as

$$b'_{ji} = \frac{b_{ji} - \min(b_{ji})}{\max(b_{ji}) - \min(b_{ji})} \tag{10}$$

For the index of the smaller the better (cost-type attribute), the standardized equation is defined as

$$b'_{ji} = \frac{\max(b_{ji}) - b_{ji}}{\max(b_{ji}) - \min(b_{ji})} \tag{11}$$

(2) Calculating entropy value E_i of the ith evaluation index.

$$E_i = -\frac{1}{\ln n}\sum_{j=1}^{m} P_{ji} \ln P_{ji} \tag{12}$$

where $P_{ji} = \frac{b'_{ji}}{\sum\limits_{j=1}^{m} b'_{ji}}$, P_{ji} is the characteristic proportion of b'_{ji} in matrix B'.

(3) Determining the objective weights.

The entropy weight $w_i^{(2)}$ of the evaluation index is obtained by normalizing entropy value E_i.

$$w_i^{(2)} = \frac{1 - E_i}{\sum\limits_{i=1}^{n}(1 - E_i)} \tag{13}$$

2.2.3. Combination Weighting

The combination weighting method of game theory aims at the Nash equilibrium. It is a coordinated and integrated process to achieve consistency and compromise among different weighting methods, which maximizes the common interests of different weighting methods [42,43]. However, the combination coefficient obtained may be negative, so the constraint condition is introduced to improve the combination weighting method of game theory [44]. The specific steps of the ICWMGT are as follows:

(1) Suppose $W^{(1)} = \left[w_1^{(1)}, w_2^{(1)}, \cdots w_n^{(1)}\right]^T$, which represents a weight vector weighted by the G1 method, and $w_i^{(1)}$ represents the ith weight of the G1 method, $(i = 1, 2, \cdots, n)$. Suppose $W^{(2)} = \left[w_1^{(2)}, w_2^{(2)}, \cdots w_n^{(2)}\right]^T$, which represents a weight vector weighted by the EW method, and $w_i^{(2)}$ represents the ith weight of the EW method, $(i = 1, 2, \cdots, n)$. Then, any linear combination of $W^{(1)}$ and $W^{(2)}$ weight vectors can be expressed as

$$W = \sum_{s=1}^{2} \alpha_s W^{(s)} \tag{14}$$

where W is all the possible combination weight vectors, $W = [w_1, w_2, \cdots w_n]^T$, α_s is a linear combination coefficient, and $\alpha_s > 0$, $W^{(s)}$ is the weight calculated by the single weighting method, $(s = 1, 2)$.

(2) To find the optimal W based on game theory—that is, to find a set of coefficients that minimizes the deviation between the combination weight vector and each single weight vector—the optimal weight coefficient is solved according to the equation as

$$\min_{\alpha_s} \left\| \sum_{s=1}^{2} \alpha_s W^{(s)} - W^{(l)} \right\|_2, \ (l = 1, 2) \tag{15}$$

(3) The optimal first-derivative condition of Equation (15) is expressed as

$$\sum_{s=1}^{2} \alpha_s (W^{(l)})^T W^{(s)} = \left(W^{(l)}\right)^T W^{(l)} \tag{16}$$

It can be seen that the traditional game theory cannot guarantee that the linear combination coefficient α_s is greater than 0. If it is negative, it cannot satisfy Equation (14). Combining Equation (15) and optimizing the game model, a new objective function is established as

$$\min_{\alpha_s} \sum_{l=1}^{2} \left\| \sum_{s=1}^{2} \alpha_s (W^{(l)})^T W^{(s)} - (W^{(l)})^T W^{(l)} \right\|_2 \tag{17}$$

(4) In order to ensure that the combination coefficient is non-negative, the ICWMGT optimization model is established by adding constraints.

$$\begin{cases} \min\limits_{\alpha_s} & \sum\limits_{l=1}^{2} \left\| \sum\limits_{s=1}^{2} \alpha_s (W^{(l)})^T W^{(s)} - (W^{(l)})^T W^{(l)} \right\|_2 \\ s.t. & \sum\limits_{s=1}^{2} \alpha_s^2 = 1, \alpha_s > 0 \end{cases} \tag{18}$$

(5) In order to solve the model, the optimization model of the Lagrange function is established, the linear combination coefficient α_s is obtained by partial derivative, and the combined weight coefficient α_s^* is obtained after normalization.

$$\alpha_s^* = \frac{\sum\limits_{s=1}^{2} (W^{(l)})^T W^{(s)}}{\sum\limits_{l=1}^{2} \sum\limits_{s=1}^{2} (W^{(l)})^T W^{(s)}} \tag{19}$$

(6) The combination weight W^* can be expressed as

$$W^* = \sum_{s=1}^{2} \alpha_s^* W^{(s)} \tag{20}$$

2.3. Fusion of DS Evidence Theory

2.3.1. DS Evidence Theory

DS evidence theory belongs to the reasoning method of uncertainty, which was put forward by A. P. Dempster in 1967 and was further developed by G. Shafer in 1976, so it is called DS evidence theory [45,46]. Compared with the traditional probability theory, DS evidence theory can better grasp the unknown and uncertainty of the problem.

Let Θ be the recognition framework of evidence theory, which represents the set composed of mutually exclusive and exhaustive propositions [47]. The ghost set of recognition frame Θ is marked as 2^Θ, corresponding to a set function m: $2^\Theta \to [0,1]$ on the identification framework, and it satisfies the following equation as

$$\begin{cases} m(\Phi) = 0 \\ \sum\limits_{A \subseteq \Theta} m(A) = 1 \end{cases} \tag{21}$$

where $m(A)$ is the BBA of the corresponding proposition A. If $m(A) > 0$, A is called the focal element of m, where A is any subset of Θ.

The Dempster fusion rule is the core of DS evidence theory. Let m_1 and m_2 be two BBA functions on the same recognition frame, and the focal elements are B and C, where

$A = B \cap C$. If the fusion result of m_1 and m_2 is $m = m_1 \oplus m_2$, $\oplus$ represents the orthogonal sum, and the Dempster fusion rule is defined as

$$m(A) = (m_1 \oplus m_2)(A) = \begin{cases} 0 & D = \Phi \\ \frac{\sum_{B \cap C = A} m_1(B) m_2(C)}{1-K} & D \neq \Phi \end{cases} \quad (22)$$

where $K = \sum_{B \cap C = \Phi} m_1(B) m_2(C)$, and K represents the degree of conflict between m_1 and m_2.

2.3.2. Fusion Algorithm Based on Matrix Analysis

Suppose n evaluation indexes evaluate an object together, the evaluation result has $\overline{m}$ kinds of results. If the fusion result is calculated by using the Dempster fusion rule, it will cause the problem of too much calculation, which leads to the difficulty of applying evidence theory to multi-information fusion. In this paper, the fusion algorithm based on matrix analysis is adopted, which can reduce the amount of calculation [48]. The BBAs can be expressed by the matrix of $n \times \overline{m}$.

$$M = \begin{bmatrix} m_{11} & m_{12} & \cdots & m_{1\overline{m}} \\ m_{21} & m_{22} & \cdots & m_{2\overline{m}} \\ \vdots & \vdots & \vdots & \vdots \\ m_{n1} & m_{n2} & \cdots & m_{n\overline{m}} \end{bmatrix} \quad (23)$$

where any element m_{ij} in the matrix M represents the BBA of the ith evaluation index for the jth evaluation result, $(i = 1, 2, \cdots, n, j = 1, 2, \cdots, \overline{m})$.

Since the evaluation index of DS evidence theory is assigned to $\overline{m}$ possible evaluation results, the sum of beliefs is 1. In this matrix, the sum of the elements of each row should satisfy the normalization condition.

$$m_{i1} + m_{i2} + \cdots + m_{i\overline{m}} = 1 \quad (24)$$

The transposition of the ith row of matrix M is multiplied by the kth row.

$$M_i^T \times M_k = \begin{bmatrix} m_{i1} & m_{i2} & \cdots & m_{i\overline{m}} \end{bmatrix}^T \begin{bmatrix} m_{k1} & m_{k2} & \cdots & m_{k\overline{m}} \end{bmatrix} \quad (25)$$

Then, a new matrix B of $\overline{m} \times \overline{m}$ is obtained.

$$B = \begin{bmatrix} m_{i1} \times m_{k1} & m_{i1} \times m_{k2} & \cdots & m_{i1} \times m_{k\overline{m}} \\ m_{i2} \times m_{k1} & m_{i2} \times m_{k2} & \cdots & m_{i2} \times m_{k\overline{m}} \\ \vdots & \vdots & \vdots & \vdots \\ m_{i\overline{m}} \times m_{k1} & m_{i\overline{m}} \times m_{k2} & \cdots & m_{i\overline{m}} \times m_{k\overline{m}} \end{bmatrix} \quad (26)$$

The elements of the diagonal of matrix B represent the accumulation of beliefs of two evaluation indexes, and the sum of other elements except the main diagonal elements is the degree of conflict of BBA functions.

$$K = \sum_{p \neq q} m_{ip} \times m_{kp} (p, q = 1, 2, \cdots, \overline{m}) \quad (27)$$

2.3.3. Acquisition of the BBA Function Based on Correlation Degree

In the matter-element theory, the correlation degree represents the degree to which an element belongs to a certain property, and it is necessary to transform the correlation degree of matter-element theory into the BBA of DS evidence theory. In order to maintain consistency between them, e^x function can be used for equivalent transfer transformation. Since the range of the BBA function of DS evidence theory is [0,1], the correlation degree needs to be normalized in order to establish a recognition framework that meets the requirements of DS evidence theory due to the BBA function. K_{ji} represents the correlation degree between the ith index and the jth grade of the evaluation object N_0, $(i = 1, 2, \cdots, n$, $j = 1, 2, \cdots, \overline{m})$. Through Equation (28), it is transformed into the BBA on the identification frame Θ.

$$m_i(\theta_j) = \frac{e^{K_{ji}}}{\sum\limits_{1 \le i \le n} e^{K_{ji}}} \tag{28}$$

Because the exponential function e^x is monotonously increasing, the greater the value of K_{ji}, the greater the $m_i(\theta_j)$, and the smaller the K_{ji}, the smaller the $m_i(\theta_j)$, and $0 \le m_i(\theta_j) \le 1$.

2.3.4. Fusion Algorithm Based on Combination Weights

Suppose there are n evaluation indexes for a certain object to be evaluated. Among them, the weight of the ith evaluation index is w_i^*, $(i = 1, 2, \cdots, n)$. Then, the weight vector of the BBAs provided by all evaluation indexes is $W^* = \left[w_1^*, w_2^*, \cdots w_n^*\right]$, and if $w_g^* = \max\left\{w_1^*, w_2^*, \cdots w_n^*\right\}$, $W^l = \left[w_1^*, w_2^*, \cdots w_n^*\right] / w_g^*$. Thus, it is determined that the relative importance of a BBA in all BBAs is defined as

$$\beta_i = \frac{w_i^*}{w_g^*} \tag{29}$$

β_i is called the quantity weight, and as a discount coefficient to adjust the BBA of each evaluation index, the adjusted equation is defined as

$$m_i^*(\theta_j) = \begin{cases} \beta_i m_i(\theta_j) & \theta_j \ne \Theta \\ 1 - \sum\limits_{1 \le j \le \overline{m}} m_i^*(\theta_j) & \theta_j = \Theta \end{cases} \tag{30}$$

The modified BBA $m_i^*(\theta_j)$ is used to form the BBA matrix, and the fusion algorithm based on matrix analysis is applied to obtain the evidence fusion result. The evaluation results of the object to be evaluated are judged according to the maximum membership degree principle.

3. Preparation Work for Water-Hazard Risk Evaluation

3.1. Constructing Evaluation Index System

The construction of a coal-mine water-hazard risk-evaluation index system should be comprehensive and scientific in order to ensure the reliability and accuracy of the evaluation results. Referring to the relevant data [31–34,49–51], the overall structure of the coal-mine water-hazard risk-evaluation index system is established. The mine hazard risk-evaluation index system is divided into a target layer, a first-level index and a second-level index. Taking coal-mine water-hazard risk evaluation as the target layer, according to the accident cause theory, the factors causing accidents can be summarized into four kinds of factors: human, machine, environment and management. That is, personnel factors, equipment factors, environmental factors and management factors are the necessary causes of accidents. According to the energy convergence theory, these four kinds of factors appearing at the same time will eventually lead to the occurrence of the accident. Combined with the characteristics of water hazards in coal mine, five aspects of these four types of factors are taken as first-level indexes, including personnel factors, drainage factors, water filling-source factors, water diversion-channel factors and management factors. Then, 21 coal-mine water-hazard influencing factors supporting first-level indexes are selected as second-level indexes to evaluate the grade of coal-mine water-hazard risk. The specific contents of the evaluation index system are shown in Table 2.

Table 2. Index system of coal-mine water-hazard risk evaluation.

Target Layer	First-Level Indexes	Second-Level Indexes
Coal-mine water-hazard risk evaluation	Personnel factors B_1	Professional skill level of personnel c_1
		Personnel training and education c_2
		Personnel "three violations" rate c_3
		Physical and psychological status of personnel c_4
	Drainage factors B_2	Surface waterproofing and drainage engineering c_5
		Waterproof coal pillar and waterproof gate c_6
		Water silos, pumps and drainage pipes c_7
		Power supply lines and equipment c_8
	Water filling-source factors B_3	Atmospheric precipitation c_9
		Surface water c_{10}
		Ground water c_{11}
		Old goaf water c_{12}
	Water diversion-channel factors B_4	Structural fault zone channel c_{13}
		Mining fracture zone passage c_{14}
		Artificial engineering passageway c_{15}
		Other types of water diversion channels c_{16}
	Management factors B_5	Standard degree of basic hydrogeological data c_{17}
		Perfection and implementation of rules and regulations c_{18}
		Inspection and maintenance management of equipment c_{19}
		Flood rectification and improvement c_{20}
		Accident emergency rescue capability c_{21}

3.2. Grading Standard for Evaluation Indexes

Using the relevant standard of coal-mine hazard risk classification [31–34], the coal-mine hazard risk is divided into five evaluation grades, namely, I, II, III, IV, and V, corresponding to lower risk, low risk, medium risk, high risk and higher risk, respectively. Different risk grades are assigned different characteristics, and at the same time, the scoring range of risk grades is provided as a reference for evaluation. The evaluation standard of each index is shown in Table 3.

Table 3. Evaluation standard.

Risk Grades	Risk Evaluation	Grade Characteristics	Score Range
I	lower risk	Safety, normal production	[90,100]
II	low risk	Improving risk defense	[80,89]
III	medium risk	Strictly monitoring the risk to avoid rising	[70,79]
IV	high risk	Suggesting carrying out rectification and reform	[60,69]
V	higher risk	Stop production immediately and take measures to reduce risk	[0,59]

3.3. Project Overview

Sangbei Coal Mine is located in the northeast of Hancheng City, Shaanxi Province, China, and the location is shown in Figure 1. The designed production capacity of the mine is 1.8 million t/a, and the hydrogeological type of the mine is complex. Groundwater aquifer rock groups in the mining area can be divided into nine aquifer rock groups according to burial conditions; these constitute the main water filling source of the mine. At present, the normal water inflow of the mine is 150 m^3/h, and the maximum water inflow is 270 m^3/h. Three-dimensional seismic interpretation of the mine displays a total of 89 fault lines and 12 collapsed columns, which constitute a natural water diversion channel. At present, the No. 3 coal seam is mined in this mine, and the average thickness is 5.47 m. The mining of the thick coal seam has caused damage to the roof and floor, forming a mining fracture, which constitutes an artificial water diversion channel. Because the floor elevation of the No. 3 coal seam is lower than that of the Ordovician limestone karst water level, it belongs to pressure mining under the Ordovician limestone karst water level. When it is located in the area of structural development and the waterproof layer of the floor is damaged by structure, or there is a vertical water diversion channel, coal seam mining may cause Ordovician limestone water outburst and the hidden danger of mine water inrush. According to the investigation, there have been water inrush accidents in adjacent mines. In 1975, Xiangshan Coal Mine experienced a water inrush of 414 m^3/h; in 1976, Sangshuping Coal Mine experienced a water inrush of 1530 m^3/h; and in 1976, Magouqu Coal Mine experienced a water inrush with a maximum water inrush of 12,000 m^3/h and an average water inflow of 5956 m^3/h. This resulted in a water inrush accident in the well.

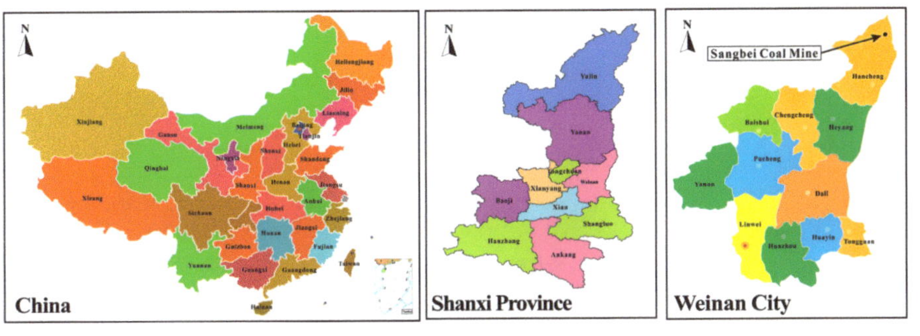

Figure 1. The location of Sangbei Coal Mine.

3.4. Determining the Value of Evaluation Indexes

Five experts from universities, design institutes and coal mines were invited to form a safety diagnosis expert group to conduct a field investigation in Sangbei Coal Mine. According to their own understanding of the actual situation of coal-mine water-hazard prevention and control, experts gave scores of 21 water-hazard evaluation indexes according to Table 3. The same weight was given to each expert, and the final scores are shown in Table 4.

Table 4. Scores of evaluation indexes.

Evaluation Indexes	c_1	c_2	c_3	c_4	c_5	c_6	c_7	c_8	c_9	c_{10}	c_{11}
Scores	81.8	82.8	86.8	87.8	87	88.6	87.4	86.6	85.4	84	82.6
Evaluation indexes	c_{12}	c_{13}	c_{14}	c_{15}	c_{16}	c_{17}	c_{18}	c_{19}	c_{20}	c_{21}	——
Scores	91.2	81.6	80.8	81	85.2	84.2	79.6	82.4	84.4	83.6	——

4. Application of Evaluation Methods

4.1. Establishing Extension Model

(1) Determining the classical domain, node domain and matter element to be evaluated.

According to Table 3 and Equation (1), the classical domains matter-element R_j related to water-hazard risk evaluation is determined, $(j = 1, 2, \cdots 5)$.

$$R_1 = \begin{bmatrix} N_1 & c_1 & [90,100] \\ & c_2 & [90,100] \\ & \vdots & \vdots \\ & c_{21} & [90,100] \end{bmatrix}, \quad R_2 = \begin{bmatrix} N_2 & c_1 & [80,89] \\ & c_2 & [80,89] \\ & \vdots & \vdots \\ & c_{21} & [80,89] \end{bmatrix}, \quad R_3 = \begin{bmatrix} N_3 & c_1 & [70,79] \\ & c_2 & [70,79] \\ & \vdots & \vdots \\ & c_{21} & [70,79] \end{bmatrix},$$

$$R_4 = \begin{bmatrix} N_4 & c_1 & [60,69] \\ & c_2 & [60,69] \\ & \vdots & \vdots \\ & c_{21} & [60,69] \end{bmatrix}, \quad R_5 = \begin{bmatrix} N_5 & c_1 & [0,59] \\ & c_2 & [0,59] \\ & \vdots & \vdots \\ & c_{21} & [0,59] \end{bmatrix},$$

The nodal domain matter-element R_p of water-hazard risk evaluation is determined by Equation (2).

$$R_p = \begin{bmatrix} N_p & c_1 & [0,100] \\ & c_2 & [0,100] \\ & \vdots & \vdots \\ & c_{21} & [0,100] \end{bmatrix}$$

The matter-element R_0 to be evaluated for water-hazard risk evaluation is determined by Equation (3).

$$R_0 = \begin{bmatrix} N_0 & c_1 & 81.8 \\ & c_2 & 82.8 \\ & \vdots & \vdots \\ & c_{21} & 83.6 \end{bmatrix}$$

(2) Calculating correlation degrees.

According to Equation (4), correlation degrees of 21 evaluation indexes with respect to five hazard risk grades are calculated, and the results are shown in Table 5.

Table 5. Evaluation index grades' correlation degrees.

Evaluation Indexes	I Grade Risk	II Grade Risk	III Grade Risk	IV Grade Risk	V Grade Risk
c_1	−0.3106	0.2000	−0.1333	−0.4129	−0.5561
c_2	−0.2951	0.3111	−0.1810	−0.4452	−0.5805
c_3	−0.1951	0.2444	−0.3714	−0.5742	−0.6780
c_4	−0.1528	0.1333	−0.4190	−0.6065	−0.7024
c_5	−0.1875	0.2222	−0.3810	−0.5806	−0.6829
c_6	−0.1094	0.0444	−0.4571	−0.6323	−0.7220
c_7	−0.1711	0.1778	−0.4000	−0.5935	−0.6927
c_8	−0.2024	0.2667	−0.3619	−0.5677	−0.6732
c_9	−0.2396	0.4000	−0.3048	−0.5290	−0.6439
c_{10}	−0.2727	0.4444	−0.2381	−0.4839	−0.6098
c_{11}	−0.2984	0.2889	−0.1714	−0.4387	−0.5756
c_{12}	0.1200	−0.2000	−0.5810	−0.7161	−0.7854
c_{13}	−0.3134	0.1778	−0.1238	−0.4065	−0.5512
c_{14}	−0.3239	0.0889	−0.0857	−0.3806	−0.5317
c_{15}	−0.3214	0.1111	−0.0952	−0.3871	−0.5366
c_{16}	−0.2449	0.4222	−0.2952	−0.5226	−0.6390
c_{17}	−0.2685	0.4667	−0.2476	−0.4903	−0.6146
c_{18}	−0.3377	−0.0192	−0.0286	−0.3419	−0.5024
c_{19}	−0.3016	0.2667	−0.1619	−0.4323	−0.5707
c_{20}	−0.2642	0.4889	−0.2571	−0.4968	−0.6195
c_{21}	−0.2807	0.4000	−0.2190	−0.4710	−0.6000

4.2. Calculating Combination Weights

The subjective weight is calculated by the G1 method. The order relation and importance ratio between indexes are obtained by the expert group's rational judgment of evaluation indexes, and the subjective weight $W^{(1)}$ is obtained by Equations (5)~(8). $w_i^{(1)}$ is shown in Table 6. Based on the scores of experts, the objective weight is calculated by the EW method, and characteristic proportion and entropy value of each index are calculated by Equations (9)~(13). The objective weight $W^{(2)}$ is obtained, and $w_i^{(2)}$ is shown in Table 6. The subjective weight $W^{(1)}$ and the objective weight $W^{(2)}$ are calculated by Equation (19) to obtain $\alpha_1^* = 0.5410$ and $\alpha_2^* = 0.4590$, and the combination weight W^* is calculated by Equation (20). The results based on ICWMGT are shown in Table 6. We can obtain $w_g^* = 0.1028$, W^l is obtained by Equation (29), and β_i is shown in Table 6.

Table 6. Weights and ranking of evaluation indexes.

Evaluation Indexes	$W^{(1)}$	$W^{(2)}$	W^*	W^l	Ranking of Importance
c_1	0.0357	0.0558	0.0449	0.4369	9
c_2	0.0297	0.0318	0.0307	0.2984	18
c_3	0.0815	0.0326	0.0591	0.5745	6
c_4	0.0153	0.0514	0.0318	0.3095	15
c_5	0.0203	0.0378	0.0283	0.2755	19
c_6	0.0270	0.0365	0.0314	0.3051	16
c_7	0.0570	0.0347	0.0468	0.4547	7
c_8	0.0518	0.0313	0.0424	0.4123	10
c_9	0.0139	0.0591	0.0346	0.3367	12
c_{10}	0.0224	0.0463	0.0333	0.3242	14
c_{11}	0.1194	0.0834	0.1028	1.0000	1
c_{12}	0.0126	0.0386	0.0245	0.2387	21
c_{13}	0.1085	0.0843	0.0974	0.9470	2
c_{14}	0.0987	0.0498	0.0762	0.7414	3
c_{15}	0.0185	0.0345	0.0259	0.2514	20
c_{16}	0.0246	0.0463	0.0346	0.3360	13
c_{17}	0.0168	0.0475	0.0309	0.3004	17
c_{18}	0.0741	0.0595	0.0674	0.6554	5
c_{19}	0.0432	0.0378	0.0407	0.3959	11
c_{20}	0.0897	0.0466	0.0699	0.6798	4
c_{21}	0.0393	0.0545	0.0463	0.4499	8

The results of ICWMGT weighting, subjective weighting and objective weighting are compared in Figure 2, and it can be seen that the fluctuation of weights obtained by the G1 weighting method is relatively large, while the fluctuation of weights obtained by the EWD weighting method is relatively small. The fluctuation of weights obtained by ICWMGT is between the two. So, ICWMGT weighting is more practical and can better reflect the impact of various indexes on coal-mine hazard risk at the same time. If the cumulative percentage of evaluation indexes is 0~80%, these are called the controlling factors [33]. Then, 14 evaluation indexes, c_1, c_3, c_7, c_8, c_9, c_{10}, c_{11}, c_{13}, c_{14}, c_{16}, c_{18}, c_{19}, c_{20} and c_{21}, account for 79.64% of the total weight. Therefore, these 14 evaluation indexes are the main controlling factors of the water-hazard risk grade of the mine, which is also the basis and starting point for the mine to carry out water-hazard prevention and control and reduce the water-hazard risk grade.

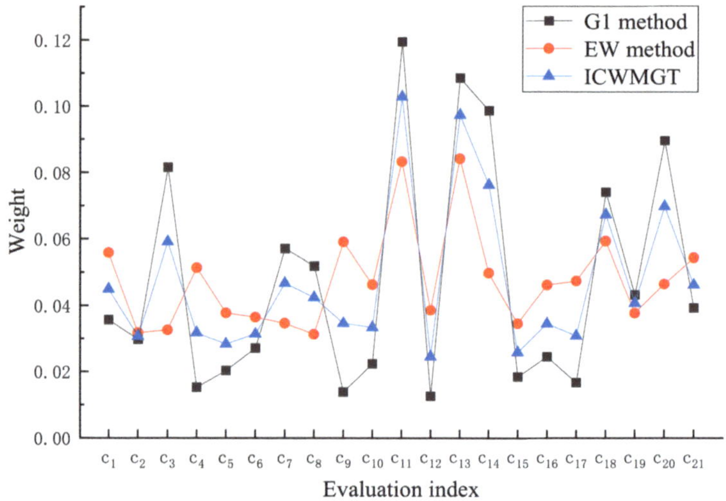

Figure 2. Comparison of the results of three weightings.

4.3. Determining Risk Grade

According to the risk classification in Table 3, coal-mine hazard-risk grade recognition framework $\Theta = \{\theta_1, \theta_2, \theta_3, \theta_4, \theta_5\}$ based on DS evidence theory is established. In the recognition framework, θ_j indicates the coal-mine water-hazard risk-evaluation grade, $(j = 1, 2, \cdots, 5)$. Then, BBA functions based on Θ are $m_1 \sim m_{21}$. Based on Table 5, using Equation (28), we can realize the transformation between the correlation degree and the BBA. Taking the quantity weight as the discount coefficient, BBAs are modified by Equation (30) based on Table 6, and then the modified BBAs of water-disaster risk evaluation are calculated as shown in Table 7.

Table 7. The modified BBAs.

Evaluation Indexes	$m_i^*(\theta_1)$	$m_i^*(\theta_2)$	$m_i^*(\theta_3)$	$m_i^*(\theta_4)$	$m_i^*(\theta_5)$	$m_i^*(\Theta)$
c_1	0.0788	0.1313	0.0941	0.0711	0.0616	0.5631
c_2	0.0536	0.0983	0.0601	0.0461	0.0403	0.7016
c_3	0.1224	0.1900	0.1027	0.0838	0.0755	0.4255
c_4	0.0718	0.0956	0.0550	0.0456	0.0414	0.6905
c_5	0.0597	0.0899	0.0492	0.0403	0.0364	0.7245
c_6	0.0761	0.0888	0.0538	0.0451	0.0413	0.6949
c_7	0.1019	0.1445	0.0811	0.0668	0.0605	0.5453
c_8	0.0864	0.1382	0.0737	0.0600	0.0540	0.5877
c_9	0.0641	0.1216	0.0601	0.0480	0.0428	0.6633
c_{10}	0.0578	0.1184	0.0599	0.0468	0.0413	0.6758
c_{11}	0.1798	0.3235	0.2042	0.1563	0.1363	0.0000
c_{12}	0.0780	0.0566	0.0387	0.0338	0.0315	0.7613
c_{13}	0.1709	0.2792	0.2065	0.1557	0.1347	0.0530
c_{14}	0.1339	0.2023	0.1699	0.1265	0.1088	0.2586
c_{15}	0.0454	0.0700	0.0569	0.0425	0.0366	0.7486
c_{16}	0.0630	0.1228	0.0599	0.0477	0.0425	0.6640
c_{17}	0.0535	0.1116	0.0546	0.0429	0.0378	0.6996
c_{18}	0.1174	0.1615	0.1600	0.1169	0.0996	0.3446
c_{19}	0.0712	0.1258	0.0819	0.0625	0.0544	0.6041
c_{20}	0.1209	0.2567	0.1217	0.0958	0.0847	0.3202
c_{21}	0.0805	0.1589	0.0856	0.0665	0.0585	0.5501

Note: $i = 1, 2, \cdots, 21$, $m_i^*(\theta_1)$ represents the BBA between the index c_i and I grade risk.

As a result, the risk grade of each evaluation index can be determined, as shown in Figure 3. It can be seen that except the index c_{12} being of the I grade risk, all the indexes are the II grade risk. It can be seen that the distribution of risk BBAs of indexes c_{11}, c_{13}, c_{20}, c_{14}, c_3, c_{18}, c_{21}, c_7, c_8, c_1, c_{19}, c_{16}, c_9, c_{10}, c_{17}, c_2, c_4, c_5, c_6, and c_{15} gradually decreased. That is, the II grade risk of c_{11} is the most prominent, which is the key index of water-hazard prevention and control, followed by other indexes.

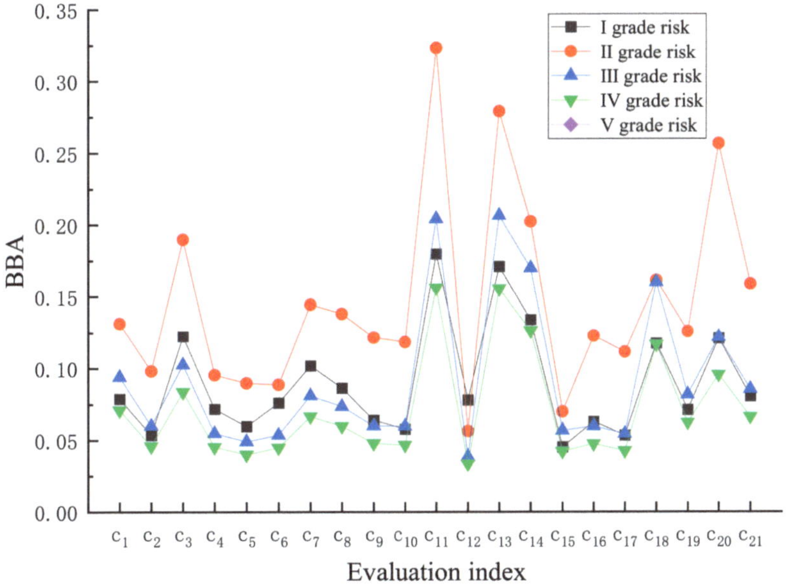

Figure 3. Risk grade of second-level indexes.

Suppose M_1^* is defined as $m_1^* \oplus m_2^* \oplus m_3^* \oplus m_4^*$, M_2^* is defined as $m_5^* \oplus m_6^* \oplus m_7^* \oplus m_8^*$, M_3^* is defined as $m_9^* \oplus m_{10}^* \oplus m_{11}^* \oplus m_{12}^*$, M_4^* is defined as $m_{13}^* \oplus m_{14}^* \oplus m_{15}^* \oplus m_{16}^*$, and M_5^* is defined as $m_{17}^* \oplus m_{18}^* \oplus m_{19}^* \oplus m_{20}^* \oplus m_{21}^*$, the BBA fusion results of the indexes of second-level are shown in Table 8.

Table 8. Fusion results of second-level indexes.

Evaluation Indexes	$M_k^*(\theta_1)$	$M_k^*(\theta_2)$	$M_k^*(\theta_3)$	$M_k^*(\theta_4)$	$M_k^*(\theta_5)$	$M_k^*(\Theta)$
B_1	0.1566	0.2767	0.1468	0.1119	0.0977	0.2104
B_2	0.1653	0.2552	0.1268	0.1017	0.0909	0.2602
B_3	0.1786	0.3658	0.1927	0.1416	0.1213	0.0000
B_4	0.1623	0.3395	0.2146	0.1421	0.1163	0.0252
B_5	0.1468	0.3584	0.1764	0.1218	0.1017	0.0949

Note: $k = 1, 2, \cdots, 5$, $M_k^*(\theta_1)$ represents the BBA between the index B_k and I grade risk.

The risk grades of first-level indexes can be determined as shown in Figure 4. It can be seen that the BBAs of first-level indexes are obtained from the BBA fusion of second-level indexes, and the maximum value of the BBAs is obviously improved, so it is easier to judge the risk grades of first-level indexes, all of which are II grade risk. The II grade risk of indexes B_3, B_4 and B_5 is relatively certain, but it can also be seen that these three first-level indexes also have potential safety risks, and there is a tendency towards the III grade risk. In particular, the hidden danger of the index B_4 is more prominent. Combined with Figure 3, we can note that among four indexes, c_{13}, c_{14} and c_{15} are supporting index B_4 and are all of II grade risk. Combined with

Table 6 for further analysis, we can see that the ranking of the impact weights of indexes c_{13} and c_{14} is second and third, respectively. Both of them are the main controlling factors of water-hazard risk. This is the main starting point and basis for improving the risk grade of the index B_4. The indexes c_{11} and c_{10}, supporting the index B_3, are also classified as II grade risk and have a tendency towards III grade risk. The impact weights of these two indexes are ranked first and fourteenth, and both belong to the main controlling factors of water-hazard risk. The impact of the index c_{11} is particularly prominent. Five indexes, c_{17}, c_{18}, c_{19}, c_{20} and c_{21}, supporting the index B_5 are all the II grade risk. Due to the ranking of the impact weights of indexes c_{20}, c_{18} and c_{21} as fourth, fifth, and eighth, respectively, these three indexes are also the main controlling factors of water-hazard risk. Therefore, in order to reduce the risk grade of indexes B_3, B_4, and B_5, further water-hazard risk-control work should be carried out on the above seven indexes first, in order to achieve significant water-hazard prevention and control effects. Although the II grade risk of indexes B_1 and B_2 are relatively certain, they have a tendency towards the I grade risk. The reason for this is that the coal mine has performed well in these two aspects and has received unanimous recognition from experts. As the index c_1 is the II grade risk and also a major control factor, the mining company should further improve the technical level of personnel, while the other three major control factors of c_3, c_7 and c_8 should continue to be maintained and continuously improved to achieve the I grade risk. Although the index c_2 is not the main controlling factor, it should also be noted by the mining company, and personnel training and education should be further improved.

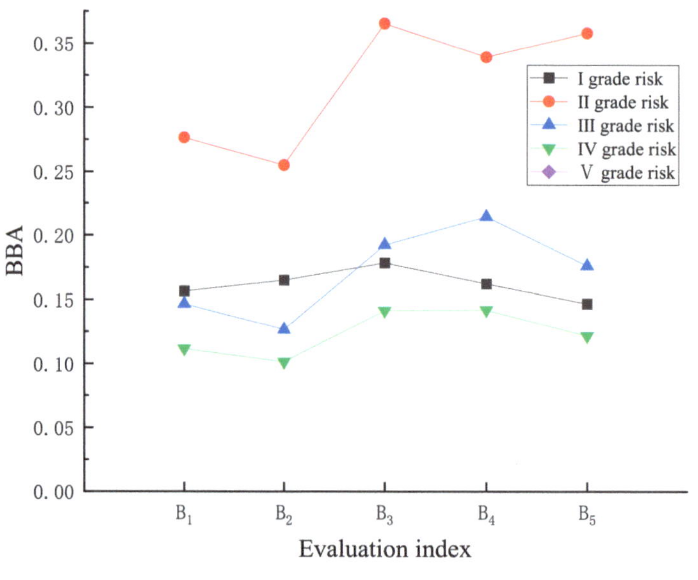

Figure 4. Risk grade of first-level indexes.

Suppose M_1^{**} is defined as $M_1^* \oplus M_2^*$, M_2^{**} is defined as $M_1^{**} \oplus M_3^*$, M_3^{**} is defined as $M_2^{**} \oplus M_4^*$, and M_4^{**} is defined as $M_3^{**} \oplus M_5^*$. Furthermore, the BBA fusion results of first-level indexes are shown in Table 9.

Table 9. Fusion results of first-level indexes.

Fusion Frequency	$M_k^{**}(\theta_1)$	$M_k^{**}(\theta_2)$	$M_k^{**}(\theta_3)$	$M_k^{**}(\theta_4)$	$M_k^{**}(\theta_5)$	$M_k^{**}(\Theta)$
1	0.1839	0.3561	0.1515	0.1122	0.0969	0.0993
2	0.1584	0.5217	0.1514	0.0938	0.0746	0.0000
3	0.1052	0.6734	0.1285	0.0556	0.0373	0.0000
4	0.0660	0.7930	0.0906	0.0313	0.0191	0.0000

Note: $k = 1, 2, \cdots, 4$, $M_k^{**}(\theta_1)$ represents the BBA for I grade risk after the kth fusion.

The maximum BBA of the fusion results is 0.7930; it can be determined that the water hazard of Sangbei Coal Mine is the II grade risk, corresponding to low risk. The investigation is in line with the current situation of water-hazard prevention and control in the mine. Although the hydrogeological conditions of Sangbei Coal Mine are complex, the current working face has not yet entered the mining area with pressure, so the risk of safety production being affected by highly confined water is low. Due to the good performance of water prevention equipment, water control management and water prevention staff in the mine, this is also one of the main reasons why Sangbei Coal Mine is the II grade water-hazard risk. Because the water diversion channel is one of the main hazard factors causing water inrush in coal mines, Sangbei Coal Mine should strengthen the detailed exploration of underground structures, accurately locate and control fault structures, study the failure law of roofs and floors, and master the development of the roof fracture zones and floor failure depths of mining faces. For water filling sources, Sangbei Coal Mine should establish and improve the water source database and strengthen the prediction of water filling sources. The prediction of water filling sources is economical and reasonable, which is also the basic premise of waterproofing work.

5. Discussion

(1) The traditional combinatorial weighting method of game theory cannot ensure that the linear combination coefficient is greater than 0. Here, the traditional game theory is used for combination weighting by Equation (16), and the coefficient calculation result is $\alpha_1 = 1.0537$, $\alpha_2 = -0.0718$. It can be seen that coefficient α_2 is negative, so it is contradictory to the hypothesis. This also greatly limits the application scope of game theory, and according to the ICWMGT, we cannot only ensure that the parameter is non-negative; rather, we must also achieve the goal of game theory. Therefore, the ICWMGT has more scientific and extensive application prospects.

(2) The BBA fusion is directly carried out by Equation (22), and the time required for calculation is $T(6^{21})$. If there is an increase in the number of indexes, the time required will increase exponentially—it can cause "focal element explosion". In order to solve this problem, a fusion algorithm based on matrix analysis can greatly improve the computational efficiency, and the whole calculation time is $T(20 \times 6^2)$. As a result, the complexity of time is reduced from the original exponential level to the polynomial level, and there is an obvious difference between them in the number of operations and amount of operation time.

(3) The extension theory can solve the problem of the incompatibility of various features of things, DS evidence theory can better deal with the combination of fuzzy and uncertain information, and the description of uncertain problems is closer to people's habits of thinking. The organic combination of extension theory and DS evidence theory can improve the accuracy of risk evaluation. Based on the combination weights obtained in this paper, the expert scoring method [31], fuzzy comprehensive evaluation method [33], and extension evaluation method [52] were used to evaluate the coal-mine hazard risk, respectively, to further verify the accuracy of the evaluation results in this paper. The expert scoring method was used to obtain an evaluation score of 83.8, which is a II grade risk. The fuzzy comprehensive evaluation method was used to obtain an evaluation score of 85.48, which is a II grade risk. The comprehen-

sive correlation degrees of each risk grade obtained using the extension method are −0.2566, 0.2362, −0.2319, −0.4797, and −0.6066, respectively. Therefore, the highest correlation degree with a II grade risk is 0.2362. All of these results are consistent with the evaluation result obtained in this paper, but it can be seen that the method used in this paper is more scientifically reasonable. The expert scoring method calculates the sum of the product of each index' score and its weight to obtain the evaluation result; although it is simple and practical, it cannot reflect the unknown uncertainty and incompatibility in the evaluation process, and it is also unable to further explore the potential intrinsic information, such as the degree of membership between various indexes and risk grades. The membership degree calculation process of the fuzzy evaluation method fails to fully utilize expert scoring results, which makes it difficult to comprehensively reflect the risk membership degree of each index, and the synthesized algorithm needs further exploration. The extension method can only solve the problem of incompatible features of things, but it cannot effectively handle uncertain information. Therefore, the reasoning process of the three methods mentioned above is not as rigorous as the method proposed in this paper.

6. Conclusions

In response to the shortcomings of current coal-mine water-hazard risk-evaluation models and methods, a risk evaluation method based on extension theory, game theory and DS evidence theory is put forward. Based on the accident cause theory, the evaluation index system and risk-evaluation grade of coal-mine water-hazard risk are established. Based on the concepts of matter elements and correlation function, the relationship between the evaluation index of coal-mine water-hazard risk and each risk-evaluation grade is quantified and normalized, so as to obtain the BBA of the evaluation index. The objectivity of obtaining the BBA is improved. This paper improves the game theory and puts forward the ICWMGT, and it can avoid the weight deviation that may be caused by the single weighting method in the process of weight calculation. Based on the combination weights, the BBAs of evaluation indexes are modified, and a fusion algorithm based on combination weights is used to fuse the modified BBAs, which can reduce the uncertainty in the evaluation process and obtain results from coal-mine hazard risk evaluation. The grade of coal-mine water-hazard risk is consistent with the actual situation of coal mines, and the evaluation results are verified by another three methods at the same time. The main conclusions in this paper are as follows:

(1) The ICWMGT was used to optimize the combination of the G1 method weighting results and the EW method weighting results, with Nash equilibrium as the coordination objective. This can balance the conflicts between them and obtain the optimal weighting combination that takes into account the characteristics of different weighting methods. This overcomes the uncertainty and one-sidedness of the single weighting method in the decision-making process and improves the scientific rationality of index weighting. According to the calculation results of ICWMGT, the importance ranking of 21 evaluation indexes had been determined, and 14 evaluation indexes could be identified as the control factors for coal-mine water-hazard risk. This was the basis and starting point for the mine to carry out water-hazard prevention and control and reduce the water-hazard risk grade.

(2) Based on extension theory, the classical domain, section domain, and evaluated object elements of coal-mine water-hazard risk evaluation were determined. The correlation degree of 21 evaluation indexes for coal-mine water-hazard risk evaluation with respect to five water-hazard risk grades was calculated according to the correlation function. The obtained comprehensive correlation degree was equivalently transformed and normalized using the e^x function to obtain the BBA on the evidence theory recognition framework. The BBA was modified by ICWMGT weighting results. Based on the fusion of DS evidence theory, the risk grades of various primary indicators were determined. Based on the weighting results of ICWMGT and the

BBAs results of secondary indexes, it was found that 11 secondary indexes were the focus of coal-mine water-hazard risk prevention and control, among which seven indexes were the primary starting point for coal-mine water-hazard risk prevention and control—especially the groundwater index, which had the most prominent impact.

(3) The water-hazard risk of Sangbei Coal Mine was the II grade, and the results were consistent with the hydrogeological conditions of the mine. Based on this result, the expert scoring method, fuzzy comprehensive evaluation method, and extension evaluation method were used to evaluate the risk of coal-mine water hazard, respectively, to further verify the accuracy of the evaluation results in this paper. Through comparison, it was found that the method used in this paper is scientifically reasonable. At the same time, the ICWMGT proposed in this paper can overcome the disadvantage of the traditional game theory weighting method that may obtain negative coefficients. The fusion algorithm based on matrix analysis can reduce the complexity of computation time from the exponential level to the polynomial level, greatly improving computational efficiency, and the description of uncertain problems is closer to human thinking habits. Thus, the method proposed in this paper has certain promotion and application value, and it can also be evaluated and applied in other fields.

Author Contributions: Conceptualization, X.X. and X.W.; methodology, X.X. and G.S.; validation, X.X. and G.S.; formal analysis, X.W.; investigation, X.X., X.W. and G.S.; resources, X.W. and G.S.; writing—original draft preparation, X.X.; writing—review and editing, X.X. and X.W. All authors have read and agreed to the published version of the manuscript.

Funding: This research was funded by the Henan Province Science and Technology Research and Development Program (232102320233) and the Key Scientific Research Projects in the Colleges and Universities of Henan Province (22A440008).

Data Availability Statement: The data are available from the corresponding author upon reasonable request.

Conflicts of Interest: The authors declare no conflicts of interest.

References

1. Yin, H.; Zhou, W.; LaMoreaux, J.W. Water inrush conceptual site models for coal mines of China. *Environ. Earth Sci.* **2018**, *77*, 746.
2. Wu, Q. Progress, problems and prospects of prevention and control technology of mine water and reutilization in China. *J. China Coal Soc.* **2014**, *39*, 795–805.
3. Cao, Z.; Gu, Q.; Huang, Z.; Fu, J. Risk assessment of fault water inrush during deep mining. *Int. J. Min. Sci. Technol.* **2022**, *32*, 423–434. [CrossRef]
4. Dong, F.; Yin, H.; Cheng, W.; Zhang, C.; Zhang, D.; Ding, H.; Lu, C.; Wang, Y. Quantitative prediction model and prewarning system of water yield capacity (WYC) from coal seam roof based on deep learning and joint advanced detection. *Energy* **2024**, *290*, 130200.
5. Bai, Z.; Liu, Q.; Liu, Y. Risk assessment of water inrush from coal seam roof with an AHP–CRITIC algorithm in Liuzhuang Coal Mine, China. *Arab. J. Geosci.* **2022**, *15*, 364.
6. Sun, W.; Li, W.; Ning, D.; Ren, L. Current states, prediction and prevention suggestions for water hazard accidents in China's coal mines. *Coal Geol. Explor.* **2023**, *51*, 185–194.
7. Bureau, N.C.M.S. *Coal Mine Water Control Rules*; China Coal Industry Publishing House: Beijing, China, 2018; p. 1.
8. Ma, D.; Duan, H.; Zhang, J.; Bai, H. A state-of-the-art review on rock seepage mechanism of water inrush disaster in coal mines. *Int. J. Coal Sci. Technol.* **2022**, *9*, 50.
9. Hu, W.; Zhao, C. Trilinear chart classification method of mine water hazard type based on factors of water recharge. *Coal Geol. Explor.* **2019**, *47*, 2.
10. Wu, Q.; Guo, X.; Bian, K.; Du, X.; Xu, K.; Bu, W.; Zeng, Y. Carrying out general survey of the water disaster-causing factors to prevent the occurrence of coal mine water disasters. *China Coal* **2023**, *49*, 3–15.
11. Dong, D.; Zhang, J. Discrimination Methods of Mine Inrush Water Source. *Water* **2023**, *15*, 3237. [CrossRef]
12. Wu, Q.; Mu, W.; Xing, Y.; Qian, C.; Shen, J.; Wang, Y.; Zhao, D. Source discrimination of mine water inrush using multiple methods: A case study from the Beiyangzhuang Mine, Northern China. *Bull. Eng. Geol. Environ.* **2019**, *78*, 469–482. [CrossRef]
13. Yan, Z.; Han, J.; Yu, J.; Yang, Y. Water inrush sources monitoring and identification based on mine IoT. *Concurr. Comput. Pract. Exp.* **2019**, *31*, e4843.
14. Jiang, Q.; Liu, Q.; Liu, Y.; Chai, H.; Zhu, J. Groundwater chemical characteristic analysis and water source identification model study in gubei coal mine, Northern Anhui Province, China. *Heliyon* **2024**, *10*, e26925. [PubMed]

15. Li, B.; Wu, Q.; Liu, Z.; Zhou, L. Identification of Mine Water Inrush Source Based on PCA-FDA: Xiandewang Coal Mine Case. *Geofluids* **2020**, *2020*, 2584094.
16. Wang, J.; Park, H.D. Coal mining above a confined aquifer. *Int. J. Rock Mech. Min.* **2003**, *40*, 537–551. [CrossRef]
17. Liu, Z.; Liu, Q.; Liu, Y. Classification of hidden faults in coal seam floor and measures of water inrush prevention. *Coal Geol. Explor.* **2020**, *48*, 141–146.
18. Wu, L.; Bai, H.; Yuan, C.; Wu, G.; Xu, C.; Du, Y.; Moayedi, H. A Water-Rock Coupled Model for Fault Water Inrush: A Case Study in Xiaochang Coal Mine, China. *Adv. Civ. Eng.* **2019**, *2019*, 9343917. [CrossRef]
19. Song, W.; Liang, Z. Theoretical and numerical investigations on mining-induced fault activation and groundwater outburst of coal seam floor. *Bull. Eng. Geol. Environ.* **2021**, *80*, 5757–5768. [CrossRef]
20. Shao, J.; Zhang, Q.; Zhang, W. Evolution of mining-induced water inrush disaster from a hidden fault in coal seam floor based on a coupled stress–seepage–damage model. *Geomech. Geophys. Geo-Energ. Geo-Resour.* **2024**, *10*, 78. [CrossRef]
21. Sun, W.; Liu, H.; Cao, Z.; Yang, H.; Li, J. Mechanism Analysis of Floor Water Inrush Based on Criteria Importance though Intercrieria Correlation. *Water* **2023**, *15*, 232. [CrossRef]
22. Duan, H.; Zhao, L. New evaluation and prediction method to determine the risk of water inrush from mining coal seam floor. *Environ. Earth Sci.* **2021**, *80*, 30. [CrossRef]
23. Hebblewhite, B. Fracturing, caving propagation and influence of mining on groundwater above longwall panels—A review of predictive models. *Int. J. Min. Sci. Technol.* **2020**, *30*, 49–54. [CrossRef]
24. Huang, W.; Sui, L.; Wang, Y.; Zhang, C.; Jiang, D.; Cai, X.; Yang, Z. Study of the mining and aquifer interactions in complex geological conditions and its management. *Sci. Rep.* **2023**, *13*, 9462. [CrossRef] [PubMed]
25. Zhang, J.; Wu, J.; Yang, T.; Yang, S.; He, Y.; Gao, S. Analysis of Water Inrush Disaster Mechanism of Inter-Layer Rocks between Close Coal Seams under the Influence of Mining. *Appl. Sci.* **2023**, *13*, 9043. [CrossRef]
26. Gu, H.; Tao, M.; Li, X.; Cao, W.; Li, Q. Dynamic response and meso-deterioration mechanism of water-saturated sandstone under different porosities. *Measurement* **2021**, *167*, 108275. [CrossRef]
27. Chen, L.; Feng, X.; Xie, W.; Xu, D. Prediction of water-inrush risk areas in process of mining under the unconsolidated and confined aquifer: A case study from the Qidong coal mine in China. *Environ. Earth Sci.* **2016**, *75*, 706. [CrossRef]
28. Huang, L.; Xu, Y.; Liu, S.; Gai, Q.; Miao, W.; Li, Y.; Zhao, L. Research on the Development Law of Pre-Mining Microseisms and Risk Assessment of Floor Water Inrush: A Case Study of the Wutongzhuang Coal Mine in China. *Sustainability* **2022**, *14*, 9774. [CrossRef]
29. Ruan, Z.; Li, C.; Wu, A.; Wang, Y. A New Risk Assessment Model for Underground Mine Water Inrush Based on AHP and D–S Evidence Theory. *Mine Water Environ.* **2019**, *38*, 488–496. [CrossRef]
30. Wang, C.; Sun, Y.; Hang, Y. Using Safety Checklist in Assessment of Potential Risk of Water—Inrush from Medium and Small Mine. *J. Min. Saf. Eng.* **2009**, *26*, 297–303.
31. Wang, P.; Li, C.; Li, Z.; Zhao, Y. Risk Assessment of Mine Water Inrush Based on Analytic Hierarchy Process. *Metal Mine* **2012**, *29*, 95–98.
32. Zhao, B. On the AHP-fuzzy comprehensive evaluation of threatening degree caused by mining water floods. *J. Saf. Environ.* **2013**, *13*, 231–234.
33. Xu, X.; Guo, B.; Tian, K.; Wang, G. Fuzzy comprehensive evaluation of coal mine water disaster risk based on combination weighting. *J. Catastrophology* **2018**, *33*, 14–18.
34. Li, Y.; Bai, J.; Yan, W.; Wang, X.; Wu, B.; Liu, S.; Xu, J.; Sun, J. Risk early warning evaluation of coal mine water inrush based on complex network and its application. *Adv. Civ. Eng.* **2021**, *2021*, 9980948. [CrossRef]
35. Chen, G.; Liu, X. Risk assessment of gushing water in coal mine based on F-ANP model. *Min. Saf. Environ. Prot.* **2023**, *50*, 129–134.
36. Tang, J.; Wang, C.; Lin, N.; Li, Z.; Li, H.; Mao, Z. Application of matter-element model in soil nutrient evaluation of ecological fragile region. *Chin. Geogr. Sci.* **2009**, *19*, 168–176. [CrossRef]
37. Cai, W. Extension theory and its application. *Chin. Sci. Bull.* **1999**, *44*, 1538–1548. [CrossRef]
38. Wang, S.; Li, L.; Cheng, S.; Liu, Z.; Ding, R.; You, Q. Model on Improved Variable Weight-Matter Element Theory for Risk Assessment of Water Inrush in Karst Tunnels. *Geotech. Geol. Eng.* **2021**, *39*, 3533–3548. [CrossRef]
39. Lu, Y.; Nie, C.; Zhou, D.; Shi, L. Research on programmatic multi-attribute decision-making problem: An example of bridge pile foundation project in karst area. *PLoS ONE* **2023**, *18*, e295296. [CrossRef]
40. Fu, P.; Zhan, Z.; Wu, C. Efficiency analysis of Chinese Road Systems with DEA and order relation analysis method: Externality concerned. *Procedia-Soc. Behav. Sci.* **2013**, *96*, 1227–1238. [CrossRef]
41. Banadkouki, M.R.Z. Selection of strategies to improve energy efficiency in industry: A hybrid approach using entropy weight method and fuzzy TOPSIS. *Energy* **2023**, *279*, 128070. [CrossRef]
42. Lai, C.; Chen, X.; Chen, X.; Wang, Z.; Wu, X.; Zhao, S. A fuzzy comprehensive evaluation model for flood risk based on the combination weight of game theory. *Nat. Hazards* **2015**, *77*, 1243–1259.
43. Peng, J.; Zhang, J. Urban flooding risk assessment based on GIS- game theory combination weight: A case study of Zhengzhou City. *Int. J. Disaster Risk Reduct.* **2022**, *77*, 103080.
44. Wu, R.; Sun, H.; Yan, D.; Tao, H.; Liao, W.; Chen, H.; Gui, D. Evaluation of flood disaster risk in China-Pakistan Economic Corridor by combination weighting based on improved game theory and grid data. *Trans. Chin. Soc. Agric. Eng.* **2021**, *37*, 145–154.
45. Dempster, A.P. Upper and Lower Probabilities Induced by a Multivalued Mapping. *Ann. Math. Stat.* **1967**, *38*, 325–339.

46. Shafer, G. *A Mathematical Theory of Evidence*; Princeton University Press: Princeton, NJ, USA, 1976; Volume 42.
47. Truong, T.; Li, Z. Incorporated Dempster-Shafer Theory, MACONT, and e-STEP Method (DSM-eSTEP) for Multicriteria Tradeoff Analysis in Transportation Budget Allocation. *IEEE Access* **2023**, *11*, 78522–78537.
48. Qu, D.; Zhang, B.; Huang, J. Application of Matrix Analysis Based DS Evidence Theory in Netted Radar. *Electron. Opt. Control.* **2010**, *17*, 77–80.
49. Gui, H.; Qiu, H.; Chen, Z.; Ding, P.; Zhao, H.; Li, J. An overview of surface water hazards in China coal mines and disaster-causing mechanism. *Arab. J. Geosci.* **2020**, *13*, 67.
50. Sun, W.; Zhou, W.; Jiao, J. Hydrogeological Classification and Water Inrush Accidents in China's Coal Mines. *Mine Water Environ.* **2016**, *35*, 214–220.
51. Wu, M.; Ye, Y.; Hu, N.; Wang, Q.; Tan, W. Visualization Analysis and Progress of Mine Water Inrush Disaster-Related Research. *Mine Water Environ.* **2022**, *41*, 599–613. [CrossRef]
52. Gong, J.; Liu, Y.; Chen, W. Land suitability evaluation for development using a matter-element model: A case study in Zengcheng, Guangzhou, China. *Land Use Policy* **2012**, *29*, 464–472. [CrossRef]

MDPI AG
Grosspeteranlage 5
4052 Basel
Switzerland
Tel.: +41 61 683 77 34

Water Editorial Office
E-mail: water@mdpi.com
www.mdpi.com/journal/water

www.ingramcontent.com/pod-product-compliance
Lightning Source LLC
Chambersburg PA
CBHW041352210926
51850CB00013B/3060
* 9 7 8 3 7 2 5 8 2 4 9 0 8 *